SpringerBriefs in Materials

The SpringerBriefs Series in Materials presents highly relevant, concise monographs on a wide range of topics covering fundamental advances and new applications in the field. Areas of interest include topical information on innovative, structural and functional materials and composites as well as fundamental principles, physical properties, materials theory and design. SpringerBriefs present succinct summaries of cutting-edge research and practical applications across a wide spectrum of fields. Featuring compact volumes of 50 to 125 pages, the series covers a range of content from professional to academic. Typical topics might include

- A timely report of state-of-the art analytical techniques
- A bridge between new research results, as published in journal articles, and a contextual literature review
- A snapshot of a hot or emerging topic
- An in-depth case study or clinical example
- A presentation of core concepts that students must understand in order to make independent contributions

Briefs are characterized by fast, global electronic dissemination, standard publishing contracts, standardized manuscript preparation and formatting guidelines, and expedited production schedules.

More information about this series at http://www.springer.com/series/10111

Kazuaki Wagatsuma

Spectroscopy for Materials Analysis

An Introduction

 Springer

Kazuaki Wagatsuma
Institute for Materials Research
Tohoku University
Sendai, Japan

ISSN 2192-1091   ISSN 2192-1105   (electronic)
SpringerBriefs in Materials
ISBN 978-981-16-5945-4   ISBN 978-981-16-5946-1   (eBook)
https://doi.org/10.1007/978-981-16-5946-1

This Springer imprint is published by the registered company Springer Nature Singapore Pte Ltd.
The registered company address is: 152 Beach Road, #21-01/04 Gateway East, Singapore 189721,
Singapore

Preface

The spectrometric methods have played an indispensable role in the material analysis which contributes to the development of modern material industry, because they can provide the analytical information on the quality/process control of the final products. Among various spectroscopic methods, X-ray fluorescence spectroscopy, electron spectroscopy, and atomic emission spectroscopy are especially employed for this purpose. Therefore, a theoretical and practical understanding of these methods is important to apply them for the actual analysis adequately. This book is written for a lecture of undergraduate students who major in the material science, which aims to explain their principles only with basic mathematical expressions and to introduce the applications to actual materials. An emphasis is placed on the interpretation for typical examples of the observed spectra using the fundamental principles of spectroscopy. It is also recommended as an introductory textbook of the recent material analysis for undergraduate/graduate students and engineers in the field of technology. I am grateful to Dr. Susumu Imashuku and Dr. Ken-ichi Nakayama (IMR, Tohoku University) for obtaining the spectrum data to prepare many figures of this book.

Sendai, Japan
June 2021

Kazuaki Wagatsuma

Contents

Chapter 1
Introduction—Material Analytical Science

1.1 Isn't ANALYSIS an Academic Discipline?

When we talk about "analysis," we think of "Analytical Chemistry" in the natural sciences; however, it is doubtful whether "analytical chemistry" is in an academic system, and it should be said that it is not. In the first place, the word "analysis" is used as a common noun in a wide range of concepts. In social sciences such as statistics, it is always mentioned in surveys and analyzes, and anyone can understand that it is an operation to stare at data and derive something from them. See Fig. 1.1. This is a rough historical overview of academic fields for "Chemistry" and several milestones that have occurred in metallic materials. It is thought that analytical chemistry has sprouted from the time of ancient metalware civilization by *identifying and classifying objects*, and then it was shined on the rise of medieval alchemy which led to discovery of the elements. However, after that, it was incorporated into the system of "inorganic chemistry", "organic chemistry" and "quantum chemistry" established in the early modern period, and the academic position of "analytical chemistry" became unclear. In short, it should be said that the methodologies and principles used in "analysis" since the Middle Ages were reaffirmed by newly developed chemistry, and there was almost nothing left in analytical chemistry.

So, "Analytical Chemistry" is no longer worth discussing? The answer is no; conversely, its importance is increasing as it supports the modern and sophisticated industrial base. However, its position is far from "chemistry" in the narrow sense, and in the manufacturing industry such as metal materials, it is closely related to engineering fields such as "metallurgy" and "electrical engineering". It has now been reorganized as a complex field that incorporates the principles and measurement methods systematized in "applied physics" and "physical chemistry". This is called "Analytical Science".

© The Author(s), under exclusive license to Springer Nature Singapore Pte Ltd. 2021
K. Wagatsuma, *Spectroscopy for Materials Analysis*, SpringerBriefs in Materials,
https://doi.org/10.1007/978-981-16-5946-1_1

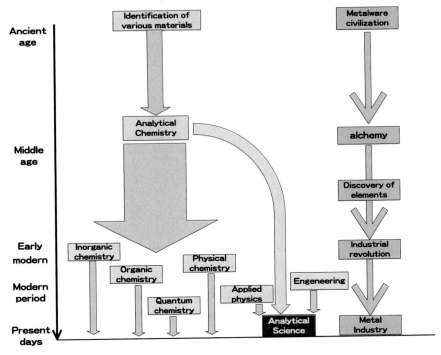

Fig. 1.1 Milestones in the development of metallic materials relating to the academic fields of chemistry

1.2 Material Analytical Science

We mentioned the newly systematized "analytical science" in the previous section; furthermore, its characteristics will be explained especially in relation to the materials industry. The analysis discussed here will be referred to as *material analysis science*. It may provide various kinds of the analytical results, but the main information is the elemental composition and its content that make up a material, that is, "Qualitative and Quantitative Analysis". Whereas, the properties required for a material are more straightforward, such as the tensile strength of the material. In general, the actual performance required for a material is not necessarily the analytical information itself; however, the usefulness of the analytical information is not lost. This is because the information obtained from the analysis can control the properties of the material.

See Fig. 1.2. This block diagram illustrates a process of developing a new material and mass-producing it. First, in developing a material, we should aim for the material properties that should be achieved. For example, the tensile strength of a material needs to be 100 kg/cm^2. The alloy/elemental composition of the material is important for obtaining the tensile strength and the other properties. There are many possibilities to try to achieve the goal based on the analytical result, which is called *development analysis*. On the other hand, *management analysis* is performed in the process of

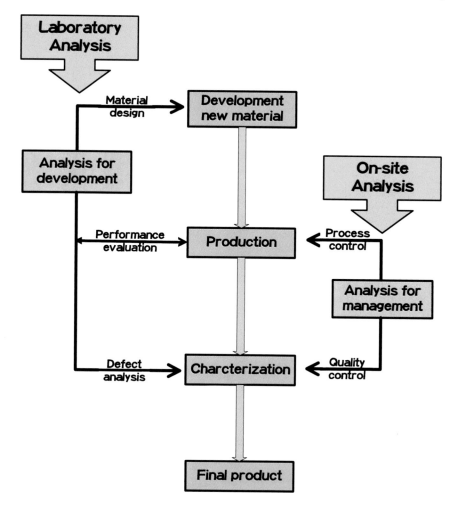

Fig. 1.2 Analytical information in the process of material production

mass production after the material design is determined. The analytical information handled by both is the same, but they are treated in a different manner because they have different purposes. The former generally evaluates small amounts of samples, but in some cases it is necessary to analyze many types of samples. The time required for the analysis is not strict, while accurate analytical results are required. On the other hand, in the latter case, the number of analysis targets is very large, and it is required the analytical values are obtained rapidly at the production process site. The main purpose is quality control of the final products. The former is sometimes called "Laboratory Analysis" and the latter "On-site Analysis". Development analysis and management analysis may use different analytical instruments.

1.3 Classification of Analytical Methods

What kind of information can be obtained by analytical methods in *material analysis science* mentioned in Sect. 1.2? First, they can be classified into several categories from their general information of analysis, as summarized in Fig. 1.3. This is a basic concept that includes all analytical chemistry, not just material analytical science. The first is classification for analytical purposes, which can be divided into "qualitative analysis", "quantitative analysis", and "state analysis". The former two are issues that have been treated by classical analytical chemistry that has continued since ancient times, and they elucidate what basic elements (atoms, compounds, etc.) are composed of substances, materials, etc., and further determine their contents. On the other hand, state analysis has become an important issue in "analytical science" in the modern era, and it is to clarify the state in which atoms, compounds, etc. are contained in substances and materials. The information from state analysis is useful in material analytical science. This is because, as higher performance of materials is required, not only the types and amounts of elements but also the physical/chemical forms of them may play significant roles in the functions of the materials.

The following item is a major classification from the analysis method. It consists of "Direct Analysis" and "Indirect Analysis". When determining what kind of elements are included in a substance, if the atoms to be analyzed can be separated independently, qualitative/quantitative analysis will be directly applied such that their weight, volume, and amount of electricity can be measured. In the method of directly measuring the physical quantity attached to an atom (this is called the *SI unit system*), the target atom has to be separated completely from other atoms. Separation/concentration operations using various chemical reactions are indispensable for this. Therefore, this method is generally called a *chemical analysis method* rather than a *direct analysis method*. On the other hand, what happens if the content

Classification of analytical methods

(1) Analytical purpose
   (a) Qualitative analysis
         determines the kind of elements in a specimen.
   (b) Quantitative analysis
         determines the content of elements in a specimen.
   (c) State analysis
         determines the physical/chemical forms of elements in a specimen.

(2) Analytical information
   (a) Direct analysis (Chemical analysis)
         measures the physical quantity directly relating to the amount of elements in a specimen, such as weight.
   (b) Indirect analysis (Instrumental analysis)
         measures the physical quantity originated from elements in a specimen, such as the intensity of radiation, which is then converted into their amounts.

Fig. 1.3 Classification of analytical methods

of the atom to be measured is very low? It will be difficult to completely separate it from other atoms, and even if it can be separated, it will be difficult to measure its amount. This is a limit of chemical analysis, and the separation ability determines the limit of quantification of analytical values. In such a case, instead of the physical quantity (for example, weight) directly attached to the atom, a method of measuring the information (for example, electromagnetic wave) emanated from the atom and then converting it on a quantitative scale is generally employed in material analytical science. This method is called "Instrumental Analysis" because it requires any measuring device that replaces the human eye. Instrumental analysis are widely conducted today as a common analytical method, since it generally requires little separation of analytes and the information obtained is highly sensitive. However, chemical analysis is still needed as a quantitative standard to link the analytical results by instrumental analysis to the SI unit system. There are various instrumental analytical methods, and also the information that can be obtained has a large variety. In the following chapters, we will focus on several instrumental analytical methods that are useful for material analysis science.

To further explain the classification of analytical methods, we will take the analysis made on SUS304 stainless steel as an example. See Fig. 1.4. The first important thing is the elemental/chemical composition that makes up this steel. Evaluations are conducted based on analytical information in the category of qualitative and quantitative analysis. This steel grade, commonly known as *18-8 stainless steel*, contains 18% chromium and 8% nickel added as the alloying elements, and its composition range is regulated by the Japanese Industrial Standard. Therefore, quantitative analysis is needed to produce the steel product that meets the standard. In addition, oxygen, nitrogen, etc., which are detected in very small amounts, are not added as alloying elements but exist as impurity elements. Because these minor elements also affect the steel properties, their analytical information is indispensable to control the impurity amounts. Furthermore, the main purpose of this steel grade is corrosion

Evaluation of 18-8 stainless steel

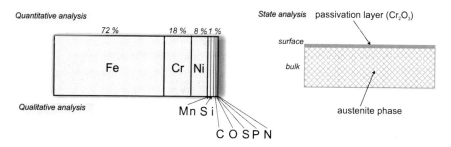

Fig. 1.4 Evaluation of 18-8 stainless steel as an example of qualitative, quantitative, and state analysis

resistance (to prevent rust and maintain surface gloss). In terms of corrosion resistance, a *passivation film* formed on the surface plays an important role; thus, state analysis methods are applied to investigate the surface properties. Also, its metallurgical structure is the austenite phase, which can be observed and identified by other instrumental analytical methods.

1.4 Laboratory Analysis and On-site Analysis

The main purpose of the laboratory analysis, which is presented in Fig. 1.2, is in line with that of "analytical chemistry". We will take an analyte sample to a laboratory equipped with analytical equipment, and after thoroughly considering what kind of analytical information is needed, a true value of the analytical result should be provided. Therefore, the precision and accuracy is primarily considered as the characteristics of the *laboratory analysis*, as illustrated in Fig. 1.5. On the other hand, the most important factor in the *on-site analysis* is the response time of analytical information. A metaphorical explanation for this relationship is customs inspection at an airport. See Fig. 1.6. Drug finder for the purpose of detecting drug smuggling, a drug-sniffing dog is active in the baggage claim area of the airport. Customs clearance requires the ability to handle large volumes of luggage. In short, detection dogs instantly determine the presence or absence of drugs on the spot. This leads to a similar role to the on-site analysis in the field of material production. The detection dogs cannot provide information about the type or amount of narcotics, so suspected

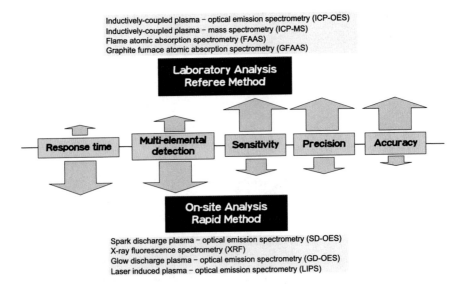

Fig. 1.5 Analytical characteristics for laboratory and on-site analysis, together with widely-employed analytical methods belonging to each category

Drug-detection dog and High-performance liquid chromatography

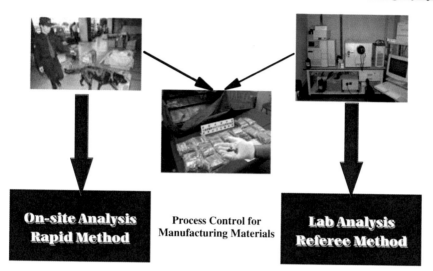

Fig. 1.6 Role of laboratory analysis and on-site analysis, as a similarity to that of the drug-detection dog and high-performance liquid chromatography at a customs site of airport

ones will be subjected to precise analysis using equipment at customs facilities. *High performance liquid chromatography* is a widely used method for qualitative and quantitative analysis of organic compounds. By conducting this measurement, the first decision of the detection dog becomes evidence-proof in the judiciary. This is exactly the role required for laboratory analysis.

Figure 1.5 also indicates commonly-employed methods for spectroscopic analysis belonging to each category. Since it is not possible to introduce these analytical methods in detail here, their names will be just denoted in the diagram, but *radiofrequency inductively coupled plasma—optical emission spectroscopy* (ICP-OES) is widely used not only in the materials industry but also in qualitative and quantitative analysis for various analytes. Whereas ICP-OES can provide highly sensitive and accurate analytical values, the analysis target has to be taken from the manufacturing site to the facility for analysis, because it requires pretreatment with a sample as a solution, On the other hand, *spark discharge optical emission spectroscopy* (SD-OES) and *fluorescent X-ray spectroscopy* (XRF) are used for on-site analysis. In the materials industry, SD-OES is a uniquely developed analytical method and is indispensable for the manufacturing sites. It has a more complex role than the drug detection dog mentioned above, but basically requires the prompt response of analytical values. It is a key analytical method for the process control and quality control.

1.5   Progress to Online Analysis

As considering the purpose of on-site analysis at the manufacturing site, that is, rapid processing of mass-produced products, it is ideal to complete the analysis procedure simultaneously while transporting the analyte objects by a moving body such as a belt conveyor. This is so-called "on-line analysis". The on-line analysis has already been put into practical use in several industrial fields. For example, in a fruit sorting process, the on-line sorting is realized not only for its size and weight, but also for the internal factors such as maturity and sugar content.

There have been some cases where on-line analysis is conducted in the field of metal materials; however, further improvements are still left for the full application. This has strict requirements imposed on the controlled analysis of metallic materials, and multiple elements with contents ranging from tens of percent to 0.001% must be quantified at the same time. Also, surprisingly important is that metal materials at the manufacturing stage are often different from the final product. In particular, it is a problem of the analysis portion, and the materials during the production process have surface stains, oxide layers, surface roughness etc., which hinder accurate analysis. This situation is quite different from the on-line fruit sorting. It should be noted that the fruit sorting is pre-processed with careful manual pretreatment, such as cleaning and sorting at the time of harvest. Compared to metal materials, fruit products are generally expensive and then have high added value, and even if a small amount of them is handled carefully, the economic rationality is not lost.

The most promising on-line analysis method for metal materials is *laser-induced plasma—optical emission spectrometry* (LIBS). *Glow discharge plasma—optical emission spectrometry* (GD-OES) is also employed at manufacturing sites in the metal industry. This article mainly deals with such analytical methods, so they will be explained in detail in Chap. 15, but here I will explain the core of why LIBS can perform on-line analysis. The point is that LIPS allows analysis without a direct contact with a sample as well as with little pretreatment of it.

Chapter 2
Concept of Spectroscopy

2.1 Excitation and De-excitation

It is mentioned in Sect. 1.3 that *instrumental analysis* is a method of measuring information emanated from atoms, etc., to the outside environment instead of the physical quantity (for example, weight) directly attached to the analyte target. We need to further understand here what kind of information is emitted by the atoms, and what kind of means should be taken to obtain the information. The contents are summarized schematically in Fig. 2.1. Atoms can emit various particles and energy waves depending on the environment. Typical examples are electrons and ions, electromagnetic waves such as X-rays and ultraviolet radiation, heat, and sound. It should be noted here that the majority of them do not have analytical information, or even if they have, they are difficult to be detected and thus cannot be effectively used in material analysis. For example, heat is a normally generated energy release, but it is not used in material analysis because it is difficult to identify the atoms in the sample, which means poor qualitative analytical information. Another point to be considered is that it is necessary to work from the external environment to force such phenomena to occur, in order that we can obtain particles and energy release emitted from the atoms. In the ambient temperature and pressure, atoms are stable and rarely emit particles or electromagnetic waves to the outside. Therefore, in order to obtain the analytical information, it is necessary to supply energy to the analyte target to induce the above phenomena. For this purpose, various types of particle beams such as electron beam, electromagnetic waves, and electric discharges, are employed for the energy supply into the atoms. This is called "Excitation" in a broad sense. Atoms donated with energy are in a certain unstable state, but they will return to a stable state so quickly. This is broadly called "De-excitation". At this time, the energy difference between the unstable and stable states is released to the outside environment. Figure 2.1 schematically illustrates these relationships. More notably, this emitted energy must have a specific value by the atom kind to be employed as the analytical information. Even if the energy can be detected, it cannot be used as an identification tool for atoms if all the values are the same.

© The Author(s), under exclusive license to Springer Nature Singapore Pte Ltd. 2021
K. Wagatsuma, *Spectroscopy for Materials Analysis*, SpringerBriefs in Materials,
https://doi.org/10.1007/978-981-16-5946-1_2

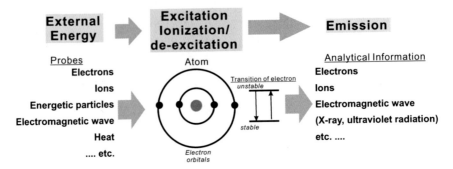

Fig. 2.1 Excitation/de-excitation phenomenon of atom and the analytical information

Now, we will see the origin of the energy release from an atom: what occurs in the atom during the excitation and de-excitation? That is, the energy difference between the unstable and stable states described above has a particular value peculiar to the atom. In electromagnetic radiation, it is attributed to the internal energy of the electron orbits associated with the atom. In other words, the radiation is emitted when an electron jumps from a higher energy electron orbital to a lower electron orbital. At this time, light in a broad sense is observed in wavelength ranges from visible, ultraviolet, to X-ray. By investigating this phenomenon, qualitative and quantitative analyses of atoms become possible. As to why the electron orbital has such a property, we have to understand a theory of quantum chemistry called "energy quantization". This will be explained briefly in the following sections.

Generally, the method of measuring electromagnetic waves from a substance is called "Spectroscopy" or "Spectrometry" especially for the analytical application. Furthermore, in a broad sense, *spectroscopy* is used not only in detection of electromagnetic waves but also in measuring methods based on the energy selection of electrons and ions. Figure 2.2 summarizes the information obtained from *spectrometry*. The electromagnetic waves emitted from an analyte target are divided by energy, which is called *energy dispersion*. They are characterized by energy units such as wavelength, frequency, and wave number. Each energy value should be accompanied by an intensity factor. Similarly, electrons and ions can be dispersed by their kinetic energies. A graph with the energy value on the X-axis and the intensity on the Y-axis is called a *spectrum* (see Fig. 2.2). The spectrum contains all the analytical information. In other words, the X-axis shows atomic species to be analyzed (qualitative analysis), and the Y-axis shows an amount of the atomic species (quantitative analysis).

Spectroscopy

measures electromagnetic radiation, electron, and ion by energy (energy dispersion).

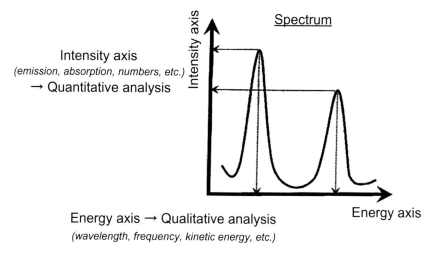

Fig. 2.2 Concept of spectroscopic measurement

2.2 Electron Orbital

In Sect. 2.1, we understand that the information obtained from spectroscopy is the electromagnetic radiation derived from the energy difference when an electron jumps from a higher-energy electron orbital to a lower-energy electron orbital. So what is an electron orbital? As an important fact for considering this phenomenon, it is necessary to recognize that *electron orbitals are discrete and have discontinuous energy*. If the orbitals are continuous (have continuous energy), the spectrum shown in Fig. 2.2 will not appear; however, this is a wrong idea because the spectrum comprises discrete spectral lines. In fact, such discontinuous energy states cannot be interpreted by classical mechanics. It is easy to understand if we recall what is happening in the world of human scale (1 m). For example, when lifting something, we can fix it at any height position (continuous potential energy). A new academic system was established by *Schrödinger's wave equation* through Bohr's early quantum theory to solve the problem of energy discreteness occurring on the electron scale, leading to quantum mechanics. By mathematically solving the wave function, the shape of an electron orbital and the energy value as its eigenvalue can be obtained, and the energy discreteness of the electron orbitals can be understood. Since the wave function is a mathematical treatment, it is difficult to indicate its physical meaning straightforwardly. However, a simple explanation is that *electrons are massive particles and at the same time they have the character of waves*.

The waves in our environment disappear fairly quickly, such as ripples that occur on the surface of the water. This is explained by waves interfering with each other

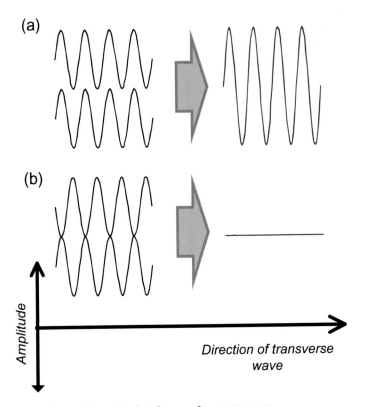

Fig. 2.3 Constructive and destructive interference of transverse wave

and then decaying. This is called *destructive interference*. Figure 2.3 shows a model of wave interferences. If you pay attention to the node positions of the waves, i.e., peaks and bottoms of the waves, you can see in Fig. 2.3b that they cancel each other and disappear instantly, which is called an *anti-phase state*. This is an extreme example, but spontaneously-generated waves are destined to decay because they normally are out of phase. The only thing that does not disappear is when the phases are perfectly matched, as shown in Fig. 2.3a. This is called *standing-wave resonance or constructive interference*. The core of the wave function is that the electron wave never disappears; otherwise, the existence probability of the electrons would become zero. We can consider the solution of the wave equation so as to satisfy this condition, leading to discrete energy states of the electron orbitals.

Figure 2.4 shows the shape of the electron orbitals obtained as a solution to the wave equation [1]. What is more important in spectroscopic analysis is the energy of the orbital rather than the shape itself. Four quantum numbers are allocated as factors for characterizing the orbitals; that is, "the principal quantum number (n)", "the azimuth quantum number (l)", "the magnetic quantum number (m)", and "the spin quantum number (s)". Then, the electron orbitals defined by these quantum numbers appear to surround the nucleus hierarchically, such as *K shell, L shell, M*

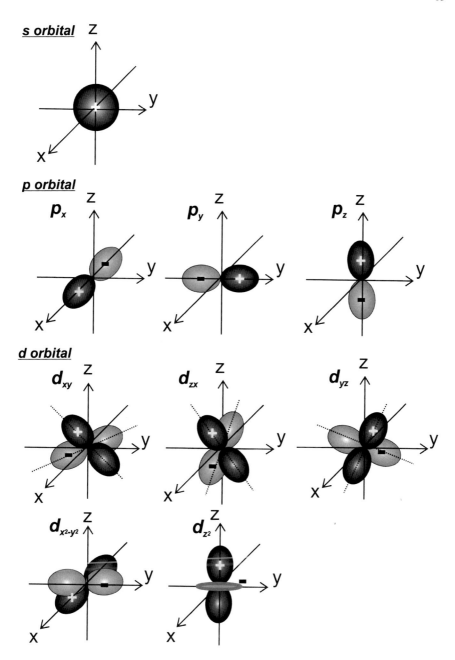

Fig. 2.4 Schematic drawing of electron orbitals

Relationship between electron orbitals and the quantum numbers

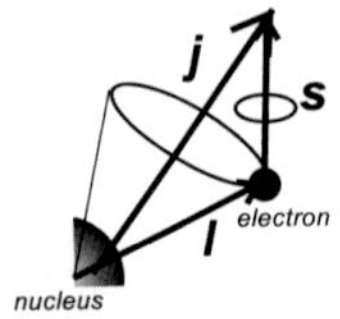

Shell	K	L				M								
n	1	2				3								
Orbital	1s	2s	2p			3s	3p			3d				
l	0	0	1			0	1			2				
m	0	0	-1	0	+1	0	-1	0	+1	-2	-1	0	+1	+2
s	±½	±½	±½	±½	±½	±½	±½	±½	±½	±½	±½	±½	±½	±½

Diagram labels: j, s, electron, nucleus

n: principle quantum number
l: azimuthal quantum number
m: magnetic quantum number
s: spin quantum number

Fig. 2.5 Quantum numbers indicating the arrangement and energy of electrons in an atom

shell, and so on, which accompany corresponding sub-shells, as indicated in Fig. 2.5. The details will not be further explained. If we think of these as just a sign, we can fully understand the discussion in the following sections. An important point is that *a combination of the quantum numbers (n, l, m, s) uniquely determines the corresponding energy value.*

2.3 Electron Configuration, Ground State and Excited State

It is easy to understand that an electron orbital located near the nucleus is more stable. In addition, *Pauli's Exclusion Principle* must be considered in determining the electronic configuration. The most stable electron orbital arrangement is determined for each atom by the principle that electrons occupy orbitals from K shell towards the outer shells. This is called a *ground state* in the electron configuration. Table 2.1 summarizes the ground state configuration of each element, and it is the most basic information in spectroscopic analysis [2]. Next, the electron configuration of magnesium will be explained as an example.

Magnesium has an atomic number of 12, and 12 electrons occupy the electron orbitals. Let us consider what the most stable electron configuration is. As found in Table 2.1, it becomes $(1s)^2(2s)^2(2p)^6(3s)^2$. The Mg atom in this state is stable (ground state) and does not provide any information to the outside world. When a certain amount of energy is given to the Mg atom, a phenomenon occurs in which electrons move between orbitals. This is called an inter-orbital transition. Basically, there are various possibilities in the electron transitions, which are constrained by the energy value supplied. It should be noted that electron configurations resulting from the transitions are always unstable compared to the *ground-state electron configuration*. This is called *excited electron configurations*. It is understood that there is only one ground state electron configuration but many excited-state electron configurations. Let us further explain several important cases in the excited state. Figure 2.6 deals

Table 2.1 Electron configuration in the periodic table of elements

Shell	K	L		M			N			
Subshell	$l=0$	$l=0$	$l=1$	$l=0$	$l=1$	$l=2$	$l=0$	$l=1$	$l=2$	$l=3$
H	$(1s)$									
He	$(1s)^2$									
Li	$(1s)^2$	$(2s)$								
Be	$(1s)^2$	$(2s)^2$								
B	$(1s)^2$	$(2s)^2$	$(2p)$							
C	$(1s)^2$	$(2s)^2$	$(2p)^2$							
N	$(1s)^2$	$(2s)^2$	$(2p)^3$							
O	$(1s)^2$	$(2s)^2$	$(2p)^4$							
F	$(1s)^2$	$(2s)^2$	$(2p)^5$							
Ne	$(1s)^2$	$(2s)^2$	$(2p)^6$							
Na	$(1s)^2$	$(2s)^2$	$(2p)^6$	$(3s)$						
Mg	$(1s)^2$	$(2s)^2$	$(2p)^6$	$(3s)^2$						
Al	$(1s)^2$	$(2s)^2$	$(2p)^6$	$(3s)^2$	$(3p)$					
Si	$(1s)^2$	$(2s)^2$	$(2p)^6$	$(3s)^2$	$(3p)^2$					
P	$(1s)^2$	$(2s)^2$	$(2p)^6$	$(3s)^2$	$(3p)^3$					
S	$(1s)^2$	$(2s)^2$	$(2p)^6$	$(3s)^2$	$(3p)^4$					
Cl	$(1s)^2$	$(2s)^2$	$(2p)^6$	$(3s)^2$	$(3p)^5$					
Ar	$(1s)^2$	$(2s)^2$	$(2p)^6$	$(3s)^2$	$(3p)^6$					
K	$(1s)^2$	$(2s)^2$	$(2p)^6$	$(3s)^2$	$(3p)^6$		$(4s)$			
Ca	$(1s)^2$	$(2s)^2$	$(2p)^6$	$(3s)^2$	$(3p)^6$		$(4s)^2$			
Sc	$(1s)^2$	$(2s)^2$	$(2p)^6$	$(3s)^2$	$(3p)^6$	$(3d)$	$(4s)^2$			
Ti	$(1s)^2$	$(2s)^2$	$(2p)^6$	$(3s)^2$	$(3p)^6$	$(3d)^2$	$(4s)^2$			
V	$(1s)^2$	$(2s)^2$	$(2p)^6$	$(3s)^2$	$(3p)^6$	$(3d)^3$	$(4s)^2$			
Cr	$(1s)^2$	$(2s)^2$	$(2p)^6$	$(3s)^2$	$(3p)^6$	$(3d)^4$	$(4s)^2$			
Mn	$(1s)^2$	$(2s)^2$	$(2p)^6$	$(3s)^2$	$(3p)^6$	$(3d)^5$	$(4s)^2$			
Fe	$(1s)^2$	$(2s)^2$	$(2p)^6$	$(3s)^2$	$(3p)^6$	$(3d)^6$	$(4s)^2$			
Co	$(1s)^2$	$(2s)^2$	$(2p)^6$	$(3s)^2$	$(3p)^6$	$(3d)^7$	$(4s)^2$			
Ni	$(1s)^2$	$(2s)^2$	$(2p)^6$	$(3s)^2$	$(3p)^6$	$(3d)^8$	$(4s)^2$			
Cu	$(1s)^2$	$(2s)^2$	$(2p)^6$	$(3s)^2$	$(3p)^6$	$(3d)^9$	$(4s)^2$			
Zn	$(1s)^2$	$(2s)^2$	$(2p)^6$	$(3s)^2$	$(3p)^6$	$(3d)^{10}$	$(4s)^2$			

with the electron configuration of magnesium as a typical example. The electrons $(3s)^2$ in the outermost shell of the Mg atom is particularly important for inter-orbital transitions because they are weakly bound with the nucleus and thus moved easily. This electron is called the *outermost electron*. When this outermost electron moves to a vacant orbital on the outside, for example, the 3p orbital, an electromagnetic wave corresponding to the energy difference between the 3p orbital and the 3s orbital is

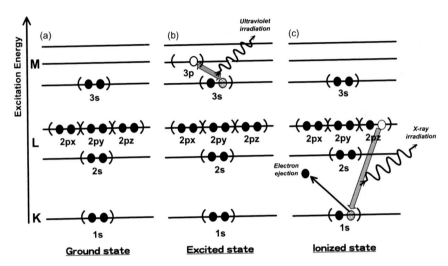

Fig. 2.6 Schematic diagram of the electron energy levels of neutral magnesium atom and typical electron transitions

emitted when it returns to the 3s orbital. This is a de-excitation process, which can be schematically drawn in Fig. 2.6b. This drawing is what is called "energy level diagram", indicating the excitation/de-excitation processes of magnesium along with the energy axis. The original energy level diagram is more complex and precise, but simplified for illustration. Excitation/de-excitation processes occur not only between the ground state electron configuration but also between the excited electron configurations. Another important excitation process occurs when a large amount of energy is supplied compared to the transition of the outermost electron. In such a case, the electrons in the inner shells, for example, electrons in the 1s orbital, can be moved. At this time, the moving electron usually jumps out of the atom, as illustrated in Fig. 2.6c. In other words, an ionization occurs and Mg^+ ion is formed; however, the electron configuration at this time is extremely unstable because it has a hole in the 1s orbital. Therefore, in order to stabilize the atom, the outer 2p orbital electron moves to fill the hole in the 1s orbital. Then, an electromagnetic wave corresponding to the energy difference between the 1s orbital and the 2p orbital is emitted. Using the diagram of Fig. 2.6, check the transitions of electrons that have been explained in this section.

References

1. Murrell JN, Kettle SFA, Trdder JM (1970) Valence theory. Wiley, London
2. Herzberg G (1944) Atomic spectra and atomic structure. Dover Publications, New York

Chapter 3
Interaction Between Electromagnetic Waves and Matter

3.1 What Is Electromagnetic Wave?

In Chap. 2, we understand that "spectrometry" is an analytical method based on the information from an *electromagnetic wave* emitted corresponding to the energy difference, when an electron in a higher energy state (orbital) jumps down to a lower energy state (orbital). Thus, it can be employed for the qualitative and quantitative measurement of the analyte target, because their energy values are peculiar for each element. So, what is the electromagnetic wave? Let us go back to the nature of electromagnetic wave and explain it in more detail.

Electromagnetic wave is a general term for traveling energy waves having transverse electric and magnetic fields that are orthogonal to each other. The electromagnetic wave is a basic constituent that governs physics, and although we cannot be conscious of it, various types of the wave are familiar to us. If there were no electromagnetic waves, the world would be only the silence of darkness. The light goes out and the mobile phone does not work. Figure 3.1 shows the wave motion of an electromagnetic wave. The magnetic field part is omitted for simplicity. Among the properties that characterize the electromagnetic wave, the energy of the wave should be emphasized in spectroscopic analysis. An extremely simple Eq. (3.1) inserted in the figure, that is, the Einstein's equation indicates that the energy is proportional to the frequency (ν) of the wave and inversely proportional to the wavelength (λ). It is an expression that is easy to be understood if we think of the electromagnetic waves as photons, and in short, it is a concept that photons with different energies are flying around us. It is correct to say that Eq. (3.1), the concept of photons, represents all the theory necessary to explain spectroscopic analysis. It's a simple theory, but it might be a proof that the natural world is governed uniquely.

By the way, how can we imagine the relationship between the wave motion of electromagnetic wave and the energy concretely? Just like the folding of an accordion curtain, the energy is high when it is fully contracted (short wavelength), and low when it is expanded (long wavelength). The wavelength range of electromagnetic waves is extremely wide, and even the waves that we can recognize range

K. Wagatsuma, *Spectroscopy for Materials Analysis*, SpringerBriefs in Materials, https://doi.org/10.1007/978-981-16-5946-1_3

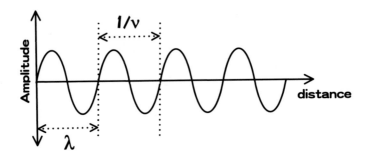

Energy of electromagnetic radiation

$$E = h\nu = hc/\lambda \qquad (3.1)$$

λ: wavelength, ν: frequency,
h: Planck cont., c: light velocity

Fig. 3.1 Nature of electromagnetic wave

in wavelengths from m order to pm (10^{-12} m). It should be noted that the photon energy is not related to the wave amplitude (fluctuation width). The amplitude of the wave determines the intensity of the electromagnetic wave, but not its energy. Therefore, there are various kinds of electromagnetic waves, such as high-energy but low-intensity waves or low-energy but high-intensity waves. Due to historical background and practical aspects, we have given various names to this extremely wide range of the electromagnetic wave. For example, electromagnetic waves that human beings recognize as light have wavelengths of 400–800 nm, blue to red visible light, and wavelengths of several nm–0.1 nm are called X-rays. Other wavelength regions have various names, but the electromagnetic waves that can be used for spectroscopic analysis are not in such a wide wavelength range. This is related to the spectroscopic information, which is outlined in Chap. 2, caused by electron jumps in an atom. Photons that have the energy required for the electron jump are mainly in visible-ultraviolet and X-ray regions.

3.2   Unit System Representing the Energy of Electromagnetic Wave

Before going on to the main issues, we will summarize the unit systems that are useful in spectroscopic analysis. As expressed in Fig. 3.1, consider the basic formula of photon energy, E, again. Since it is an amount of energy, the unit should be joule, erg, cal, etc.; however, in order to show the energy of an electromagnetic wave with a wavelength of several nm, these quantity scales are not employed directly because they need to have a power of 10. In this respect, if an *electron volt* (eV) is used, the energy of the electromagnetic wave concerned will be on the order of 1000 eV or less, and it will be possible to read and write it in an easy manner. 1 eV is a physical quantity defined by the kinetic energy which is obtained by an electron when it is placed in an electric field of 1 V, and although it is not directly related to electromagnetic waves, it is widely used in spectroscopic analysis because of the simple expression. As shown in Eq. (3.1), the physical measure that can be directly recognized for an electromagnetic wave is the wavelength value. Therefore, it is convenient to convert the wavelength into a unit of eV directly, including Planck's constant and the speed of light. 1 eV = 8065 cm^{-1} can be applied as this conversion constant. The unit of cm^{-1} is the reciprocal of centimeter, which is called *Kaiser*. Remember this conversion relationship as it is very convenient.

3.3 Interaction Between Electromagnetic Wave (X-Ray) and Substance

What happens when a solid sample surface is exposed to X-rays? Figure 3.2 schematically shows this situation. Here, it is important that an X-ray beam, having an energy of around 1000 eV, should be irradiated to the sample. If the electromagnetic wave has a long wavelength (low energy), the phenomenon will be completely different. Since the X-ray have enough energy to cause electronic transitions between the electron orbitals described in Sect. 2.3, several emission phenomena can be observed. *Photoelectrons* and *Auger electrons*, which are derived from electrons in the inner

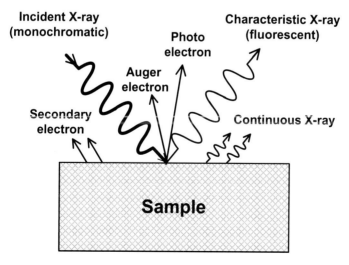

Fig. 3.2 Irradiation of X-ray on a solid surface, which induces emissions of electrons and electromagnetic waves

electron shells of the sample atom ejected by the irradiation energy of X-ray, is a powerful source of information in spectroscopic analysis. We will discuss the electron emission in Chap. 7. In addition, X-rays having various wavelengths other than the incident X-ray are emitted from the sample. This is also due to the electronic transition between electron orbitals described in Sect. 2.3, and several of the X-rays are observed at specific wavelength positions depending on the kind of sample atom. This is called "Characteristic X-ray". It is also called *fluorescent X-ray* because it is induced by the incident of an X-ray. (*Light generated by incident light is generally called fluorescence.*) The characteristic X-rays appear at particular wavelengths depending on the sample atom, so they can be employed for qualitative and quantitative analysis. The next chapter explains this principle in detail. In addition, unlike characteristic X-rays, there is X-ray radiation that appears continuously in a certain wavelength range rather than at a specific wavelength position. This is called "Continuous X-ray". *Bremsstrahlung* is a major mechanism for generating the continuous X-ray, and when the ejected electrons are de-accelerated in the sample surface, their kinetic energy is converted to X-ray emission. In addition, X-rays for which the incident X-ray itself reflects or scatters on the sample surface, radiation emitted in wavelength regions other than X-rays, heat, sound wave, etc. are observed. They are not usually applied for spectroscopic analysis because they provide poor information for identifying atomic species compared to the characteristic X-ray.

Chapter 4
Principle of X-Ray Fluorescence Spectroscopy

4.1 Binding Energy of Electron Orbital

By evaluating an electron jump (transition) between electron orbitals quantitatively, we can concretely understand the electron emission and the generation of a characteristic X-ray which are induced by the irradiation of X-ray. Here, an important physical quantity is the *binding energy of electron* for each electron orbital. It is defined by the energy required to remove an electron from a particular orbital and release it from the nucleus constraints of that orbit. In other words, it separates the electron away from the outermost vacuum level (Fermi level) of the atomic system. As already explained in Sect. 2.2, each atom accommodates electrons from the innermost electron shell (K shell) according to the number of electrons, and has a ground state in the electron configuration. At this time, the type of electron orbitals is common to all atoms, but their binding energies are different. How would we explain the reason? This is because even if the shape of the electron orbital is the same, the number of protons in the nucleus differs for each atom (the atomic number) and the number of electrons around the removed electron is also different. The binding energy of each electron orbital is different for each element, since the electron is bound under a peculiar interaction with the other electrons as well as the nucleus. Therefore, it is an important physical quantity that identifies the type of atom. The binding energy of the electron orbital has been known accurately for each atom, as listed in Table 4.1 [1]. The energy levels of each orbital in this table will be interpreted in detail in Chap. 5.

Let us check the fact that an electronic transition related to an inner electron shell corresponds to the radiation in the X-ray region. For example, the binding energy of Mg K shell (1s orbital) is 1303 eV from this table. To convert this to the wavelength of the electromagnetic wave, use the conversion formula, $1 \text{ eV} = 8065 \text{ cm}^{-1}$, as shown in Sect. 3.2. The calculation is $1303 \times 8065 = 1.05 \times 10^7 \text{ cm}^{-1}$. This is the reciprocal of the wavelength (cm); so if we take the reciprocal of this value, it will become the wavelength: $1/1.05 \times 10^7 = 0.95 \times 10^{-7} \text{ cm} = 0.95 \text{ nm}$. 0.95 nm is exactly in the wavelength range of X-rays.

© The Author(s), under exclusive license to Springer Nature Singapore Pte Ltd. 2021
K. Wagatsuma, *Spectroscopy for Materials Analysis*, SpringerBriefs in Materials,
https://doi.org/10.1007/978-981-16-5946-1_4

Table 4.1 Electron binding energy in each orbital of elements

Element/eV	$1s_{1/2}$ K	$2s_{1/2}$ L_1	$2p_{1/2}$ L_2	$2p_{3/2}$ L_3	$3s_{1/2}$ M_1	$3p_{1/2}$ M_2	$3p_{3/2}$ M_3	$3d_{3/2}$ M_4	$3d_{5/2}$ M_5
H	14								
He	26								
Li	55								
Be	111								
B	188		5						
C	284		7						
N	284		9						
O	284	24		7					
F	696	31		9					
Ne	867	45		18					
Na	1072	63		31	1				
Mg	1305	89		52	2				
Al	1560	118	74	73	1				
Si	1839	149	100	99	8		3		
P	2149	189	136	135	16		10		
S	2472	229	165	164	16		8		
Cl	2823	270	202	200	18		7		
Ar	3203	320	247	245	25		12		
K	3608	377	297	294	34		18		
Ca	4038	438	350	347	44		26		8
Sc	4493	500	407	402	54		32		7
Ti	4965	564	461	455	59		34		3
V	5465	628	520	513	66		33		2
Cr	5939	695	584	575	74		43		2
Mn	6539	769	652	641	84		49		4
Fe	7114	846	723	710	95		56		6
Co	7709	926	794	779	101		60		3
Ni	8333	1008	872	855	112		68		4
Cu	8979	1096	951	931	120		74		2
Zn	9659	1194	1044	1021	137		87		9

The numerical values are extracted from a data summary in [1]

4.2   Electronic Transition Process Related to Emission of Characteristic X-Ray

See Fig. 4.1. These are equations that express the phenomenon occurring when an electron in the K shell is ejected outside the atom by irradiation of an X-ray beam having the energy in Eq. (4.1). Equation (4.2) denotes the kinetic energy of the electron jumping out of the K shell, and is basically determined by the difference between the energy of the incident X-ray and the binding energy of the K shell electron. The electron emitted at this time is especially called "Photoelectron". Of course, this phenomenon does not occur if the energy of the incident X-ray is less than the binding energy of the K-shell. Immediately after the photoelectron is emitted, there remains a hole in the electron orbital of the K shell, which makes it extremely unstable for the atom. Various phenomena may happen to the atom to return to the stable state. This is called a *relaxation process*. As a typical relaxation process, an electron in the outer L_2-shell fills the hole of the K-shell, which can cause de-excitation in these electron orbitals. Equation (4.3) shows a mathematical expression for this de-excitation process. In this case, we observe a characteristic X-ray with the energy corresponding to the difference in binding energies between K-shell and L_2-shell. Actual spectra of the characteristic X-ray will be explained in more detail in Chap. 5. Furthermore, it should be noted that there is another important point in these equations. On the right side of Eqs. (4.2) and (4.3), we can see that correction factors are attached as the third and the higher terms. They have smaller values compared

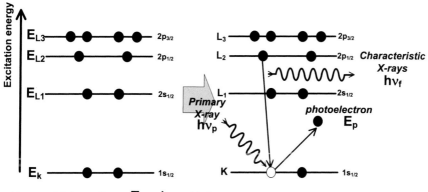

Energy of Primary X-ray: $E_{ex} = h\nu_P$ (4.1)

When an electron in K shell is ejected by the x-ray irradiation,

Energy of the photoelectron: $E_p = E_{ex} - E_K - f(K) + R_{in} + R_{ex} - \Phi$ (4.2)

When an electron in L_2 shell is de-excited into the hole of the K shell,

Energy of the characteristic X-ray: $h\nu_f = E_K - E_{L2} - f(L_2) + R_{in} + R_{ex}$ (4.3)

ν_p, ν_f : frequency of primary and characteristic X-ray. $f(x)$: correction energy for a hole of x-orbital. R_{in}, R_{ex} : intra- and inter-atomic relaxation energy. Φ : work function. h : Planck's constant

Fig. 4.1 Excitation/de-excitation processes for generating characteristic X-ray

to the first and second terms representing the energy difference; however, they are not negligibly small. In short, the energy of emitted electrons and photons cannot be determined by a simple calculation based on the binding energies. The last term at the right-hand side of Eq. (4.2) is a correction factor, Φ, which is usually called the *work function*. While, in the case of solids, the binding energy of electron orbitals is measured from the upper level (Fermi level), more kinetic energy is needed to emit the electron from the solid surface into vacuum (free electron) when all electrons are arranged in the stable state. This energy is called the work function and is usually several eV. It is not simply expressed enough to be summarized in a table but it depends on the type of substance and the state of the surface. Keep in mind that it requires extra energy equivalent to the work function to cause an electron to be ejected from a solid surface. The third term on the right side of Eqs. (4.2) and (4.3) is usually called the *correction energy for electron vacancy*. When calculating the energy difference, the binding energy is each defined by the energy value for which an electron is pulled from the perfectly-packed electron configuration, which corresponds to the ground-state configuration. What is the electron configuration when an electron falls down from the L shell to the K shell? It can be seen that the initial state of the electronic transition is an electron configuration with a hole in the K shell, which is different from that in estimation of the binding energy. Thus, the third term has to be added to correct this difference. The correction energy for vacancy is usually a few eV, but it is complicated because it also depends on the chemical/physical state of the substance. Next, the 4th and 5th terms on the right side of each equation are called *intra-atomic and inter-atomic relaxation energy*, to correct for slight variations in the binding energies of the K and L shell orbitals due to the influence by chemically-bound atoms surrounding the target atom. This value is usually on the order of 0.1 eV; however, such a correction may provide important analytical information because it is sensitive to the chemical environment of the atom, such as a metallic state or any compound. This will be explained again in relevant parts of the following chapters.

Reference

1. Someno D, Yasomori I (1976) Surface analysis. Kodansha Scientific, Tokyo

Chapter 5
Spectrum in X-Ray Fluorescence Spectroscopy

5.1 X-Ray Energy Level

In Chap. 4, it is repeatedly mentioned that an electron jump between electron orbitals may cause X-ray emission. This is an almost accurate representation; to be perfect, we need to introduce the concept of energy levels. This relates to a physical phenomenon that is also called *the splitting of spectral term*; that is, the start and end points of an electronic transition are not the electron orbital itself but several energy levels derived from an orbital (*spectral term*). See Fig. 5.1. This is a diagram showing the energy levels related to the characteristic X-ray, sometimes called "X-ray energy level". Spectral term splitting needs to be explained in terms of quantum chemistry. First, let us return to the mechanism of characteristic X-ray generation. An electron jumps out of an inner electron orbital and then the hole is filled with an electron of the outer orbital, as shown in Fig. 4.1. Therefore, if there are no electrons in the outer electron orbital, this mechanism is impossible. The typical example is a hydrogen atom, and since there is only one electron in the 1s orbital, characteristic X-rays cannot be generated. On the contrary, in the case of heavy elements, electrons are filled up to the outer electron orbitals, so various paths are created for the generation of characteristic X-rays. Now go back to the explanation of the X-ray energy level. Figure 5.2 illustrates a configuration in which the 2p orbital of the L shell is fully filled with electrons. In this electron configuration, the direction of electron spin is represented by using an arrow. The electrons form spin pairs in each of the three p-orbitals according to Pauli Exclusion Principle. This state is stable, these orbitals are energetically equivalent (*they are degraded each other*), and then the energy levels do not split as a whole. What happens if one of the electrons is removed by X-ray irradiation? The result is the same for all six electrons. One electron with no partner remains in any one of the 2p orbital. This is called an *unpaired electron*. Each electron has a spin angular momentum; however, it cancels each other out when making an electron pair and thus the momentum becomes zero. However, with an unpaired electron left, there remains the spin angular momentum which affects the energy state of the atom. In this case, the energy state splits due to the interaction

© The Author(s), under exclusive license to Springer Nature Singapore Pte Ltd. 2021
K. Wagatsuma, *Spectroscopy for Materials Analysis*, SpringerBriefs in Materials,
https://doi.org/10.1007/978-981-16-5946-1_5

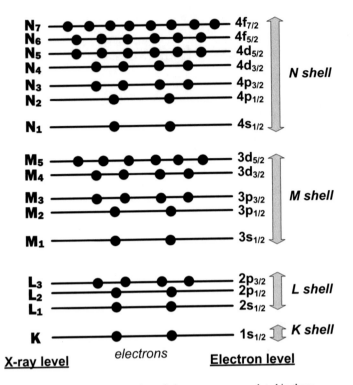

Fig. 5.1 X-ray energy levels and the number of electrons accommodated in them

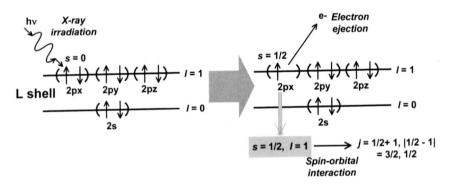

Fig. 5.2 Spin-orbital interaction for an electron hole in 2p orbitals caused by X-ray irradiation

with the orbital angular momentum of the p-orbital to which this electron belongs. The energy splitting can be explained by considering the coupling of each quantum number, that is, the spin quantum number (s) and the azimuthal quantum number (l). The coupling of the spin and orbital orientations is allowed for specific orientations according to the concept of "energy quantization". In the above example, the total angular momentum of the atom system is expressed by its quantum number (j) as inserted in Fig. 5.2: $j = (l + s), (l + s - 1), \ldots, |l - s|$, so $j = (1 + 1/2), (1 - 1/2) = 3/2, 1/2$. An energy level is obtained for each j, which determines the whole energy of the system. Thus, an electron hole in the p-orbital results in two energy levels, and the result is the same if the spin quantum number is $-1/2$. The X-ray energy levels shown in Fig. 5.1 can be understood in this way. For example, two X-ray energy levels, $2p_{1/2}$ and $2p_{3/2}$, where the j value is denoted as a subscript, are created from an electron hole in the p-orbital of the L shell. These energy levels are also labeled as L_2 and L_3, respectively.

5.2 Selection Rule

Based on the electron transition between X-ray energy levels that generates characteristic X-ray, we are able to describe the spectrum. A typical path is an electron transition from the L_2 ($2p_{1/2}$) to the K ($1s_{1/2}$) energy levels causing a major characteristic X-ray, as illustrated in Fig. 4.1. It can be understood that the spectrum is attributed to various transition paths such that electrons in the outer electron shells fill electron vacancies in the inner electron shells. At this time, there is an important rule to be further considered. That is, all of the possible transition paths are not allowed in the actual transitions of electrons. This rule is described as a physical phenomenon of *allowed or forbidden*, called "Selection rule". Figure 5.3 summarizes the selection

Selection rule for electron transition between inner orbitals

(1) Principal quantum number (n) $\Delta n \neq 0$
Any electron transition is forbidden between energy levels in the same shell.

(Example)
$2s_{1/2}$ level $\leftrightarrow 2p_{1/2,3/2}$ levels : forbidden, $3p_{1/2,3/2}$ levels $\leftrightarrow 3d_{3/2,5/2}$ levels : forbidden,
$1s_{1/2}$ level $\leftrightarrow 2p_{1/2,3/2}$ levels : allowed, $2p_{1/2,3/2}$ levels $\leftrightarrow 3d_{3/2}$ level : allowed

(2) Azimuthal quantum number (*l*) $\Delta l = \pm 1$
Change in the azimuthal quantum number must be -1, +1 in an electron transition.

(Example)
$1s_{1/2}$ level $\leftrightarrow 2s_{1/2}$ level: forbidden, $2s_{1/2}$ level $\leftrightarrow 3d_{3/2,5/2}$ levels: forbidden,
$1s_{1/2}$ level $\leftrightarrow 3p_{1/2,3/2}$ levels: allowed, $2p_{3/2}$ level $\leftrightarrow 4d_{3/2,5/2}$ levels: allowed

(3) Total quantum number (j) $\Delta j = \pm 1, 0$
Change in the total quantum number must be -1, 0, or +1 in an electron transition.

(Example)
$2p_{1/2}$ level $\leftrightarrow 3d_{5/2}$ level: forbidden, $3d_{3/2}$ level $\leftrightarrow 4f_{7/2}$ level: forbidden,
$2p_{3/2}$ level $\leftrightarrow 3d_{3/2,5/2}$ levels: allowed, $3d_{5/2}$ level $\leftrightarrow 4f_{5/2,7/2}$ levels: allowed

Fig. 5.3 Selection rule in an electron transition between inner orbitals

rules for characteristic X-ray emission. It is considered to be a restriction on the quantum numbers of the energy levels corresponding to the start and end states of a certain electron transition. For example, when an electron hole is in the 1s orbital, resulting in an X-ray energy level of $1s_{1/2}$ labeled as K, and then it will be filled with any electron in orbitals of the outer shells. However, a transition from the 2s orbital in the L shell, that is, the $2s_{1/2}$ electron (L_1) is strictly forbidden, because Δl becomes 0 which does not follow the selection rule: $\Delta l = \pm 1$. On the other hand, transitions from the 2p orbitals of the L shell, labeled as the $2p_{1/2}$ (L_2) or $2p_{3/2}$ (L_3), are allowed. It is difficult to explain the selection rule qualitatively; briefly, it is determined by the symmetry of electron orbitals in which an electron transition is involved. Electrons cannot move directly between orbitals having the same parity. Further explanations on the selection rule should be obtained in any advanced references [1].

5.3   Overview of Fluorescent X-Ray Spectrum

It is primarily important for X-ray fluorescence spectrum that the energy of the incident X-ray is greater than the binding energy of an electron in the inner shell. This can ionize an electron from the target electron orbital, and then another electron in any of the outer shells occupies the electron hole, resulting in a characteristic X-ray which corresponds to the transition path and its energy difference. Figure 5.4 summarizes an overall picture, indicating terminology to distinguish the characteristic X-rays. These were named in historical background and are somewhat confusing. Of particular importance is the mechanism in which electrons in the innermost K shell are ejected and the vacant position is filled with electrons in the outer electron shells. It is generically called *the K line*. It is possible according to the selection rule that an electron hole of the 1s orbital in the K shell is occupied by electrons in both energy levels, $2p_{1/2}$ (L_2) and $2p_{3/2}$ (L_3), of the 2p orbital in the L shell, named as *the $K\alpha_1$ and $K\alpha_2$ lines*. Similar electronic transitions are also possible from the 3p orbital in the M shell, and their characteristic X-rays are called *the $K\beta_1$ and $K\beta_2$ lines*. What happens if a 2s orbital electron in the L shell is ionized? Electrons from the p-orbitals in the outer M and N shells fill in the electron hole. Characteristic X-rays generated from a mechanism in which the vacant positions of electron are in orbitals of the L shell are collectively called *the L line*. Since the L shell also has a 2p orbital, there are also transition paths relating to this orbital and they are named as referred to Fig. 5.4. We should note that even if these transition paths are theoretically possible, the corresponding X-rays cannot be observed if there are no electrons in the orbital. Therefore, the $K\alpha_1$ and $K\alpha_2$ lines are principally observed in lighter elements. On the other hand, because heavy elements have a large number of protons in the nucleus, the K-shell electrons are trapped by high binding energies, making their ionization difficult. At this situation, we will use the L lines for analytical application, instead of the K line.

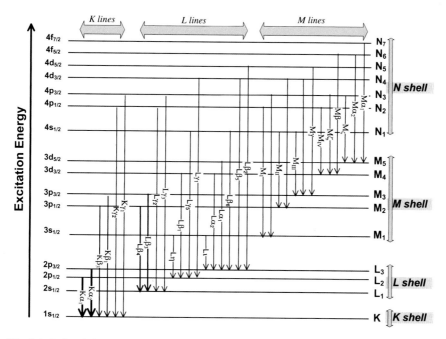

Fig. 5.4 Paths of electron transition to cause characteristic X-ray radiation

Reference

1. Corney A (1977) Atomic and laser spectroscopy. Clarendon Press, Oxford

Chapter 6
Application of X-Ray Fluorescence Spectroscopy

6.1 Wavelength Table

The method of applying characteristic X-rays to analysis exactly follows the concept of spectroscopy explained in Sect. 2.1. In other words, qualitative analysis can be carried out from the wavelength of a measured spectrum, and quantitative analysis from the intensity. The wavelength position of characteristic X-rays has been accurately measured for each element, and the database of them is compiled into "Wavelength table". Table 6.1 is a partial wavelength table of K and L lines [1], expressed in a wavelength unit of Ångstrom (Å). Ångstrom is not in the SI unit system, but it is a unit that has been widely used along with the historical development of spectroscopy, and it is converted in the relation of $1 \text{ Å} = 10^{-10} \text{ m} = 0.1 \text{ nm}$.

As an example, check the characteristic X-ray of Mg from the wavelength value in Table 6.1. A wavelength value of 9.890 Å as the $K\alpha_1$ and $K\alpha_2$ lines, and a wavelength value of 9.5207 Å as the $K\beta_1$ are found in this table. What through the transition paths do these characteristic X-rays generate? As already explained in Sect. 4.1, they are determined by the difference in the binding energies of the electron configuration between the initial and final states of the electronic transition. It can be considered by using Eq. (4.3) in Fig. 4.1, shown in Sect. 4.2. The physical meaning of the $K\alpha_1$ and $K\alpha_2$ lines of Mg is that an electron vacancy is formed in the K shell and then filled with electrons in the 2p orbital in the L shell for which the energy level is split into the $2p_{1/2}$ and $2p_{3/2}$. The values of each binding energy are found in Table 4.1, as shown in Sect. 4.1:

Mg (1s) = 1305 eV, Mg ($2p_{1/2}$) = Mg ($2p_{3/2}$) = 52 eV.

Therefore, the $K\alpha_1$ and $K\alpha_2$ lines of Mg should have the same energy as follows: Mg $K\alpha_1$, $K\alpha_2 = 1305 - 52 = 1253$ eV.

It is converted into wavelength, and it becomes $1/(1253 \times 8065) = 9.8947 \times 10^{-8}$ cm $= 9.8947$ Å, which is almost the same as the measured value of 9.890 Å listed in Table 6.1 [1]. The slight difference between these values is not due to a measurement error. The above calculation ignores the correction terms explained in Sect. 4.1, especially the vacancy relaxation energy. If it is put into the formula,

© The Author(s), under exclusive license to Springer Nature Singapore Pte Ltd. 2021
K. Wagatsuma, *Spectroscopy for Materials Analysis*, SpringerBriefs in Materials,
https://doi.org/10.1007/978-981-16-5946-1_6

Table 6.1 Wavelength table of characteristic X-ray by elements (unit: Ångstrom (Å))

Element	Kα2	Kα1	Kβ3	Kβ1	Kβ5	Kβ2	Lα2	Lα1	Lβ4	Lβ3	Lβ1
Å	$L_2 \to K$	$L_3 \to K$	$M_2 \to K$	$M_3 \to K$	$M_{4,5} \to K$	$N_{2,3} \to K$	$M_4 \to L_3$	$M_5 \to L_3$	$M_2 \to L_1$	$M_3 \to L_1$	$M_4 \to L_2$
Li		228.0									
Be		114.0									
B		67.6									
C		44.7									
N		31.6									
O		23.62									
F		18.32									
Ne		14.610		14.452							
Na		11.910		11.575							
Mg		9.890		9.5207							
Al	8.3417	8.3393		7.9605							
Si	7.1279	7.1254		6.7530							
P	6.1598	6.1568		5.7960							
S	5.3750	5.3722		5.0316							
Cl	4.7307	4.7278		4.4034							
Ar	4.1947	4.1918		3.8860							
K	3.7445	3.7414		3.4539	3.4413						
Ca	3.3617	3.3584		3.0897	3.0746						
Sc	3.0342	3.0309		2.7796	2.7634						
Ti	2.7522	2.7485		2.5139	2.4985						

(continued)

Table 6.1 (continued)

Element	Kα2	Kα1	Kβ3	Kβ1	Kβ5	Kβ2	Lα2	Lα1	Lβ4	Lβ3	Lβ1
Å	$L_2 \rightarrow K$	$L_3 \rightarrow K$	$M_2 \rightarrow K$	$M_3 \rightarrow K$	$M_{4,5} \rightarrow K$	$N_{2,3} \rightarrow K$	$M_4 \rightarrow L_3$	$M_5 \rightarrow L_3$	$M_2 \rightarrow L_1$	$M_3 \rightarrow L_1$	$M_4 \rightarrow L_2$
V	2.5074	2.5036		2.2844	2.2695						
Cr	2.2936	2.2897		2.0849	2.0709						
Mn	2.1058	2.1018		1.9102	1.891						
Fe	1.9400	1.9360		1.7566	1.7442			17.59		15.70	17.26
Co	1.7929	1.7890		1.6208	1.6089			15.972		14.27	15.666
Ni	1.6617	1.6579		1.5001	1.4886			14.561		13.16	14.271
Cu	1.5444	1.5406		1.3922	1.3816	1.3811		13.336		12.095	13.053
Zn	1.4390	1.4352		1.2953	1.2848	1.2837		12.254		11.192	11.963
Ga	1.3440	1.3401	1.2084	1.2079	1.1981	1.1960		11.292		10.365	11.023
Ge	1.2580	1.2541	1.1294	1.1289	1.1195	1.1169		10.436	9.640	9.581	10.175
As	1.1799	1.1759	1.0578	1.0573	1.0488	1.0450		9.6709		8.929	9.4141

The numerical values are extracted from a data book [1]

it should be closer to the measured value. Furthermore, the wavelength table of Table 6.1 shows a wavelength value of Mg $K\beta_1$ line. The physical meaning of the $K\beta_1$ line is the characteristic X-ray generated when an electron hole in the K shell is occupied by electrons in the 3p orbital of the M shell. However, in the ground-state electron configuration of Mg atom, this orbital is empty, so the characteristic X-ray should not be observed. However, since magnesium is actually in the form of a solid crystal or a compound, there may be electrons in this 3p orbital or in the corresponding energy level. Although the binding energy of the Mg 3p orbital is not specified in Table 4.1, it can be assumed to be a small value of ca 0.7 eV from the wavelength of the $K\beta_1$ line (9.5207 Å).

6.2   Fluorescent X-Ray Spectrum of Iron

As a typical example, Fig. 6.1 shows a measured fluorescent X-ray spectrum of iron in the K-line region. Two peaks are observed in the spectrum, known as the $K\alpha_1 (K\alpha_2)$ and $K\beta_1$ lines. What kind of the electron transition causes each characteristic X-ray? Consider their wavelength values from the binding energy of the energy levels of iron and the binding energy table, as referred to Table 4.1 in Sect. 4.1. In this spectrum, the $K\alpha_1$ and $K\alpha_2$ lines cannot be separately observed due to a limitation of the instrumental resolution power.

Fig. 6.1 X-ray fluorescent spectrum of iron by photon energy. Primary X-ray: Rh Kα, analyzing crystal: LiF(200)

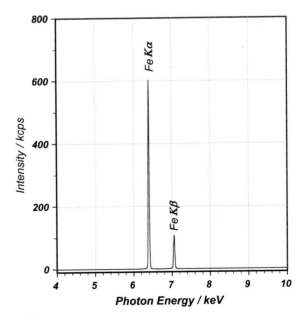

6.3 Wavelength Dispersion Measurement

How can the X-ray fluorescence spectrum of iron in Fig. 6.1 be measured? Whereas the X-ray radiation can be treated as the emission of photons, X-rays have the property of transverse electromagnetic wave. As already explained in Sect. 2.2, the progress of the transverse wave requires matching of the phase of wave; namely, the valleys and peaks of the wave are exactly in the same positions, respectively. This is called *constructive interference*. However, in natural conditions, this phase does not match and then the wave will decay and disappear, which is called *destructive interference*. By utilizing this phenomenon, it is possible to obtain an optical apparatus that selects only an X-ray line, having a specific wavelength, from emitted radiation in a certain wavelength region, especially called *wavelength dispersion*. This is a spectrometer in a broad sense.

Figure 6.2 illustrates the principle of the wavelength dispersion. When surfaces that reflect electromagnetic waves are arranged in parallel at a constant interval, the condition under which constructive interference can be achieved is determined by the surface spacing (d), the angle of incidence (θ), and the wavelength of the electromagnetic wave (λ). Equation (6.1) in this figure is called *Bragg's diffraction equation*, where electromagnetic wave with a specific wavelength can be detected by

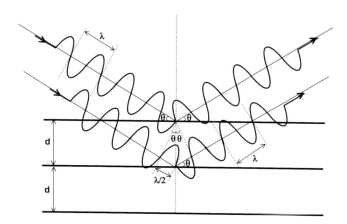

Condition for constructive interference: Bragg's equation

$$\sin \theta = n\lambda \,/\, 2d \qquad (6.1)$$

θ: diffraction angle, λ: wavelength, d: lattice distance of analyzing crystal, n: diffraction order

Fig. 6.2 Principle of wavelength-dispersed measurement for X-ray spectrum using an analyzing crystal

adjusting the surface spacing and the incident angle according to this equation. The important thing is what substances can be employed to obtain the reflective surface. As we can see from the Bragg's equation, effective results cannot be achieved when the surface spacing is wide or narrow compared to the wavelength of X-ray to be dispersed. In other words, it is necessary to prepare a substance with a surface spacing similar to the wavelength. Since the wavelength of X-rays is several 10–0.1 Å, it is difficult to artificially produce such substances with appropriate narrow intervals; however, the plane spacing of some single crystals in a specific direction, named as the *lattice constant*, is on the order of Å, and then various single-crystal materials with appropriate lattice constants as well as low attenuation to X-rays are available to be handled. This is called a *disperse/analyzing crystal*. Examples of commonly-used disperse crystals are summarized in an inserted table of Fig. 6.5. As described above, when X-rays to be measured are incident on an appropriate crystal, only a specific X-ray line appears at a certain wavelength and is observed at a specific angle, called the *diffraction angle*.

Figure 6.3 is a practical example of the X-ray fluorescence spectrum of iron in the K-ray region. While this spectrum is the same as Fig. 6.1, note that the x-axis is the diffraction angle (θ). As is clear from the Bragg's equation, the diffraction angle and the wavelength value are directly correlated; therefore, the spectrum is meaningful without any need to convert it to a wavelength value. It is common to obtain such a spectrum in this relationship of diffraction angle and intensity, as shown in Fig. 6.3, in actual measurements. In this case, it is important to specify the kind of the analyzing crystal used for the measurement, corresponding to LiF(200) having 2d = 4.02 Å indicated in this figure.

Fig. 6.3 X-ray fluorescent spectrum of iron by diffraction angle. Primary X-ray: Rh Kα, analyzing crystal: LiF(200)

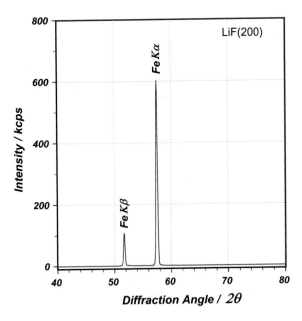

Let us explain the iron spectrum in Fig. 6.3 in more detail. Two X-ray lines appear at diffraction angles of about 52 and 58 in this spectrum. The $K\alpha_1$, $K\alpha_2$ and $K\beta_1$ lines of iron are attributed to transitions between the X-ray energy levels: $E(K\alpha_1) = 7114 - 710 = 6404$ eV, $E(K\alpha_2) = 7114 - 723 = 6391$ eV, and $E(K\beta_1) = 7114 - 56 = 7058$ eV when their binding energies are found in Table 4.1 in Sect. 4.1. Converting these energy values to wavelength: $\lambda(K\alpha_1) = 1/(6404 \times 8065) = 1.9362$ Å, $\lambda(K\alpha_2) = 1/(6391 \times 8065) = 1.9401$ Å, and $\lambda(K\beta_1) = 1/(7058 \times 8065) = 1.7568$ Å. Note that these calculated values almost correspond to the measured ones shown in Table 6.1, even though the correction terms in Eq. (4.3) are ignored. At what diffraction angles do these wavelengths appear as each X-ray line? From the Bragg's equation (Eq. 6.1), their diffraction angles are

$K\alpha_1$: $\sin \theta = n\lambda/2d = 1.9362/4.02$, $\theta = 28.79°$, $K\alpha_2$: $\sin \theta = n\lambda/2d = 1.9401/4.02$, $\theta = 28.86°$,

$K\beta_1$: $\sin \theta = n\lambda/2d = 1.7568/4.02$, $\theta = 25.91°$.

We should notice that the observed spectrum of Fig. 6.3 is measured with the sum of the incident angle and the diffraction angle, 2θ. Therefore, the $K\alpha_1$ line appears at $57.58°$ and the $K\alpha_2$ line at $57.72°$, and the $K\beta_1$ line at $51.82°$, which almost agree to the measured values. In this spectrum, two $K\alpha$ peaks cannot be separated because their wavelengths are close to each other.

The above calculation is based on the case of $n = 1$ in the Bragg's equation, where n is a positive integer called the *diffraction order*, and, in the theoretical consideration, X-ray peaks should appear even at $n = 2, 3 \ldots$ because the condition of constructive interference is satisfied. However, a mathematical condition, $-1 \leq \sin \theta \leq 1$, must restrict the range of the diffraction angle. Figure 6.4 is a practical example of the X-ray fluorescence spectrum of molybdenum in the K-ray region. In this spectrum, the $K\alpha$ and $K\beta$ lines appear at a diffraction angle of 20.6 and $18.5°$, respectively, as the primary line ($n = 1$); in addition, their second-order lines ($n = 2$) are also observed at diffraction angles that are twice as much as the first-order angles. The $K\beta_1$ and $K\beta_2$ lines can be resolved in the second order, because the resolution power becomes double when $n = 2$ in the Bragg's equation.

6.4 Quantitative Analysis

X-ray fluorescent spectrometry, which is usually abbreviated as XRF, now becomes a powerful tool for quantification of elements in various materials. The X-ray intensity does not correspond directly to the content of an analyte element in a sample but may be changed by the co-existing elements and the physical properties of the specimen, such as the surface smoothness, the metallurgical structure, and the particle-size distribution. Therefore, we need to correct for the measured intensity of the characteristic X-ray before the quantification. Figure 6.5 schematically illustrates the generation path of a characteristic X-ray having a wavelength of λ_a of the analyte element. The emitted X-ray may be absorbed by the same kind of atoms as the analyte element at positions closer to the surface (resonance *absorption*) and, at the same time, may

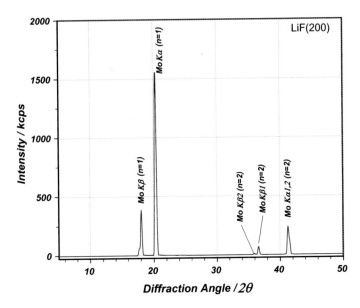

Fig. 6.4 X-ray fluorescent spectrum of molybdenum in the K region. Primary X-ray: Rh Kα, analyzing crystal: LiF(200)

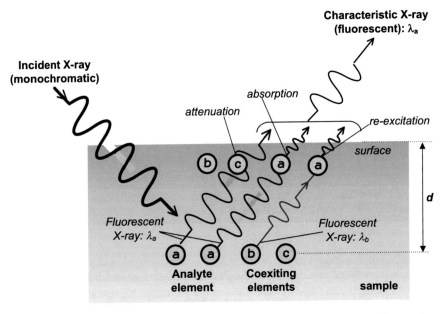

Fig. 6.5 Generation of characteristic X-rays in a sample of A-B-C ternary alloy. A is an analyte element emitting the characteristic X-ray of the wavelength, λ_a, and B is a co-existing element emitting the X-ray of λ_b ($\lambda_a > \lambda_b$)

be scattered inelastically as well as elastically and then attenuated during the progression of wave in the sample. On the other hand, if a co-exciting element is excited by the incident X-ray and generates its characteristic X-ray having a wavelength of λ_b and, further, if the wavelength is shorter than that of the analyte element (λ_a), it has the ability for exciting the characteristic X-ray of the analyte element (*re-excitation*), leading to some increase in the X-ray intensity. This re-excitation possibly occurs in a complicated manner through multiple excitation paths which are dependent on the chemical composition of the sample. These phenomena are called "Matrix Effect". The matrix effect can be corrected by a theoretical calculation, known as *the fundamental parameter method* (FP), which requires many parameters such as the incident and take-off angles of X-rays, the intensity of the incident X-ray, the wavelength of the analytical X-ray, and the chemical composition of the sample, and so on [2]. The FP software is now installed in commercial apparatuses of XRF. However, the FP correction is not enough to provide a precise analytical result in XRF, because the X-ray intensity is also affected by the state of samples.

A calibration method using a glass bead has been employed widely for the analytical applications of XRF. This is a pretreatment procedure that the sample is fused with large amounts of an X-ray transparent flux like lithium tetraborate and then the analyte atoms are homogeneously diluted, such that they are held in a clear solvent without any interactions. As a result, matrix effects derived from the state of sample can be minimized to obtain a reliable analytical result. Figure 6.6 indicates a typical example of a calibration curve in the glass bead method [3]. The target is the determination of tungsten in high-alloyed steel samples. Chemical analysis of tungsten in steel samples generally requires a time-consuming and complicated procedure for the decomposition; on the other hand, the XRF analysis can be carried out without the wet chemical operation. A linear relationship is obtained between the intensity

Fig. 6.6 Calibration relationship between the La line intensity of tungsten and the added amount of tungsten in glass bead samples including 1.1-mg steel sample and 4.0-g flux of lithium tetraborate. This article is reprinted from [3] under the permission of The Iron and Steel Institute of Japan

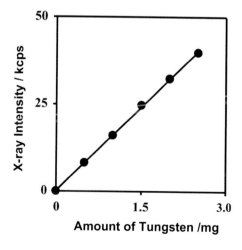

of the Lα line of tungsten and the amount of tungsten in glass beads as the quantification standard, thus contributing to the accurate determination of tungsten in steel samples.

6.5 Equipment for X-Ray Fluorescence Spectrometry

The measuring device consists of a primary X-ray source for exciting the fluorescent X-ray radiation from a sample atom, and of a spectrometer for dispersing the emitted X-ray by wavelength. Figure 6.7 is a typical example of the spectrometer, which has a rotation mechanism that can continuously change the angle of incidence on an analyzing crystal to select an arbitrary wavelength, according to the Bragg's diffraction relationship. Then, the dispersed X-ray intensity is measured with a detector. Since it is difficult to cover the wide wavelength range of X-rays with a single analyzing crystal, a method of exchanging with several ones appropriate for the wavelength ranges is adopted. The primary X-ray source is also called an *X-ray tube*. The mechanism is principally the same as the generation of the fluorescent X-ray described in this chapter. However, the X-ray tube generates an X-ray beam of a specific element by accelerating an electron beam towards the target material, instead of irradiating the X-rays, which utilizes ionization through the electron bombardment. Figure 6.8 shows a schematic diagram of the X-ray tube. The arrangement of a vacuum-sealed electron tube is used, and the anode is made of an element from which the characteristic X-ray is emanated as a primary X-ray source. When a

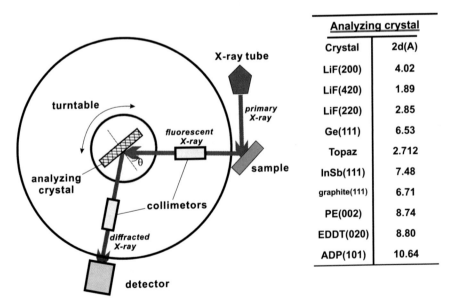

Analyzing crystal	
Crystal	2d(A)
LiF(200)	4.02
LiF(420)	1.89
LiF(220)	2.85
Ge(111)	6.53
Topaz	2.712
InSb(111)	7.48
graphite(111)	6.71
PE(002)	8.74
EDDT(020)	8.80
ADP(101)	10.64

Fig. 6.7 Schematic drawing of a spectrometer to be employed in X-ray fluorescent analysis

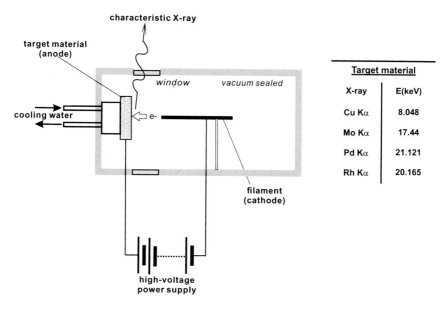

Target material

X-ray	E(keV)
Cu Kα	8.048
Mo Kα	17.44
Pd Kα	21.121
Rh Kα	20.165

Fig. 6.8 Structure of an X-ray tube to be employed as a primary source for X-ray fluorescent spectrometry

voltage of several tens of kV is applied between the electrodes, high-speed electrons collide with the anode material, and then the characteristic X-ray is emitted. Metals such as Cu, Mo, Pd, and Rh are used as the target material, and they work as a primary X-ray source depending on the wavelength of each Kα X-ray. Finally, several types of the detector are now employed for detection of the X-ray. One of the detectors, as shown in Fig. 6.9, is called a *scintillation counter (photomultiplier tube)*, which comprises a scintillation plate, a photocathode plate, and an arrangement of electron electrodes. When X-rays are incident, the scintillation material induces fluorescent radiation in a ultraviolet/visible wavelength range, and then it generates electrons due to photoelectric effect in the photocathode material. Then, the electrons are amplified through a series of the electron electrode, called a *dynode*, in which a high voltage

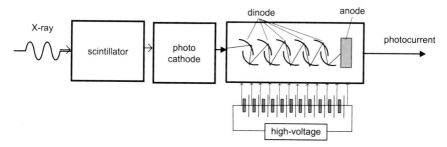

Fig. 6.9 Structure of a scintillation counter to measure the intensity of X-ray

of about 1000 V is divided and applied to each electrode, and finally detected as a photocurrent signal that is proportional to the X-ray intensity.

References

1. Bearden JA (1964) X-ray wavelengths. U.S. Atomic Energy Commission Division of Technical Information Extension, Oak Ridge
2. Shiraiwa T, Fujina N (1966) Jpn J Appl Phys 5:886–898
3. Nakayama K, Wagatsuma K (2015) Anal Sci 31:851–854

Chapter 7
Interaction Between Electrons and Matter

7.1 What Happens by Irradiating Electron Beam?

Chapters 3–6 indicate that, when a sample substance is irradiated with X-rays, the qualitative/quantitative analysis is performed based on the spectrometric information such as fluorescent X-ray and photoelectron. First, we review the role of the incident X-ray in such analysis. As repeatedly explained (see Fig. 4.1), the incident X-ray ejects an electron from an inner-shell orbital of the target atom by providing the energy of radiation larger than the binding energy, resulting in an electron hole in the orbital which is then filled with an electron from an outer-shell orbital. This process generates a characteristic X-ray. Therefore, if there is an external source that can give the energy other than the incident X-ray, it should be possible to cause a similar process to the fluorescent X-ray radiation. An electron beam can be used as the source of external energy, where the kinetic energy of the electron can be provided. In this chapter and next Chap. 8, it will be explained how qualitative/quantitative analysis is carried out using electrons that are ejected from a sample when irradiated with an electron beam.

In this respect, we have a question that a characteristic X-ray should be excited by the electron beam, if high-energy electrons have the same function as the X-ray irradiation. The answer is yes; that is, the characteristic X-ray may be also generated by the electron beam and provides the analytical information in a similar manner to fluorescent X-ray spectrometry. This is sometimes called *electron-excited characteristic X-ray*. The excitation mechanism of this characteristic X-ray is the same as that of the fluorescent X-ray, so we need not repeat it. A common analytical method using the electron-excited characteristic X-ray is *an energy dispersive X-ray spectrometer attached to a scanning electron microscope*, which is called SEM–EDX. It is widely used for material analysis because it gives information on the elemental distribution from the characteristic X-rays, in addition to microscopic observation of a sample surface. On the other hand, in the fluorescent X-ray analysis, the X-ray irradiation can also induce an electron emission from the sample, which is the principle for *X-ray photoelectron spectrometry*, as already explained using Eq. (4.2) in Sect. 4.1. In this

© The Author(s), under exclusive license to Springer Nature Singapore Pte Ltd. 2021
K. Wagatsuma, *Spectroscopy for Materials Analysis*, SpringerBriefs in Materials,
https://doi.org/10.1007/978-981-16-5946-1_7

chapter, another emission process of electrons, called "Auger electron", is explained, when high-energy electrons irradiate on a sample surface, in which an electron hole is created in the inner-shell orbital of the sample atom and then filled with electrons in the outer-shell orbitals. The analytical information is similar to fluorescent X-ray spectrometry, and is available especially for "Surface Analysis".

7.2   Difference in Excitation with Electron Beam and X-Ray Irradiation

In the previous section, we understand that an electron beam and an incident X-ray play the same role in providing the energy required for ionization of a sample. On the other hand, the interaction with a solid substance is quite different between electron and X-ray. This is essentially because the X-ray is the electromagnetic wave that can penetrate into the solid substance to a certain extent, whereas the electron beam is a flow of particles and has a negative charge. As a whole, the former has a weaker interaction with matter, while the latter has a much stronger interaction. Figure 7.1 schematically shows electron emission observed when a monochromatic electron beam (having a single kinetic energy) is incident on a sample surface, which involves several kinds of electrons. In addition, characteristic X-rays, other continuous X-rays, visible/ultraviolet radiation, heat, etc. are also generated at the same time. As only focused on these electrons, some of the incident electrons are scattered

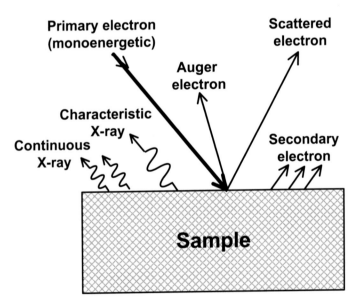

Fig. 7.1 Ejection of electrons and radiation from a solid surface when monoenergetic electron beam hits on it

and reflected on the surface with little loss of the energy. On the other hand, electrons that penetrate into the sample surface lose the kinetic energy rapidly because of the strong interaction with sample atoms, and are taken into the electronic structure of the sample and disappear. A major reaction at this time is the ionization of the sample atom, and the produced electrons may ionize other sample atoms through cascade collisions; then, their kinetic energies become significantly lower than that of the incident electron. As a result, we observe several groups of the emitted electrons having different distribution of the kinetic energy. They are called *secondary electrons*. In addition, there appear electrons with a specific kinetic energy, which corresponds to the difference in the binding energy between the inner and the outer electron orbitals of the sample atom. A typical channel for such electronic transitions is "Auger Transition", called *Auger electron*. The Auger electron is utilized for qualitative/quantitative analysis, as similar to characteristic X-rays.

Figure 7.2 shows the distribution of electron kinetic energy obtained when a metallic sample is irradiated with an electron beam having a kinetic energy of 2 keV. It is seen that the majority of electrons are secondary electrons having kinetic energies of several tens of eV and are elastically scattered electrons having the same energy as the incident electron beam. Although the number is much smaller, Auger electrons are also observed in this distribution. While the secondary electrons are measured with high intensity, they cannot be used for the analytical application because they lose most of their elemental information. On the other hand, Auger electrons have

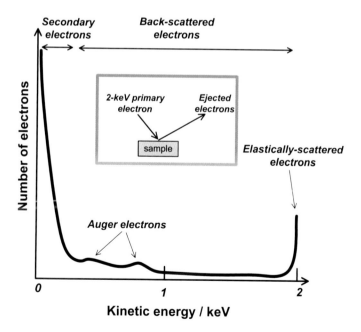

Fig. 7.2 Energy distribution of electrons emitted from a metal surface by electron bombardment of 2 keV (schematic drawing)

lower intensity, but like characteristic X-rays, they retain information on the sample atom itself as described later. There are a large difference in the ionization between an electron beam and an incident X-ray. Since the electron beam is a flux of electrons with negative charge, it can be easily accelerated in a certain electric field which is varied by the applied voltage; therefore, the kinetic energy of the electron beam can be controlled with an appropriate focusing. It is generally possible to generate an electron beam with a kinetic energy of several tens of keV that is easily varied. On the other hand, what about the X-ray source? As explained in Sect. 6.4, one has to select an appropriate characteristic X-ray from a limited number of elements. Therefore, the choice of the X-ray source is limited compared to the electron beam. However, the ionization by electrons has a disadvantage that Auger electrons have to be selected from a large number of secondary electrons, which provides a higher background in electron spectroscopy. To compensate for this difficulty, the Auger spectrum in a differential form was obtained in conventional spectrometer systems.

Chapter 8
Principle of Auger Electron Spectroscopy

8.1 Mechanism Involving Three Electrons

The excitation process of an Auger electron is schematically indicated in Fig. 8.1, involving an Auger transition labeled as KL_2L_3. The K and the L shells and their electron orbitals, and the X-ray levels have already been explained in Sect. 5.1. Strictly speaking, they exist in gaseous atoms (free atoms) and is identical to the corresponding shell orbital; however, the binding energies of the outer electron orbitals may change a little in metal crystals and compounds, because their electrons relate to any chemical bonding as a state of the outer orbitals. At this time, it should be noted that one electron is emitted from the atom in the final step and its kinetic energy is the energy difference of each orbital.

Let us explain the KL_2L_3 Auger process in more detail. First, one electron in the K shell is ionized, resulting in an electron hole. The state is very unstable because the inner shell has an electron vacancy. Therefore, any electron from the electron energy level (L_2) of the L shell fills the vacant seat of the K shell. This is the first relaxation process relating to the second electron. Even after this relaxation process, there still remains excess energy which is first generated by the ionization of the K shell. This excess energy may cause any electron in another energy level (L_3) of the L shell to be removed from the atom. The second relaxation process accompanies emission of the third electron. In this way, it is a mechanism that involves three electrons. How does this Auger process compare to the fluorescent X-ray process described in Chap. 4? It is exactly the same up to the first relaxation process, where a hole are generated in the K shell and then is filled with an electron in the L shell. In the case of X-ray fluorescence generation, the remaining energy is used for a characteristic X-ray; in other words, the second relaxation process is terminated by the X-ray emission. After the first common relaxation process, both Auger electron and characteristic X-ray are simultaneously emitted. It is a competitive relationship, but which one is superior is theoretically and experimentally clarified. In general, Auger electrons predominate for light atoms, while characteristic X-rays predominate for atoms with large atomic numbers. Equation (8.2) in Fig. 8.1 is the relational equation for calculating the

K. Wagatsuma, *Spectroscopy for Materials Analysis*, SpringerBriefs in Materials, https://doi.org/10.1007/978-981-16-5946-1_8

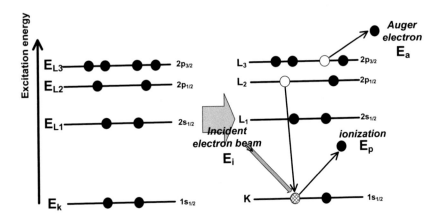

When incident electrons have an energy of E_i, the ionization occurs if $E_i \rangle E_K$ (8.1)

When an electron in K shell jumps out and then an electron in L_2 shell is de-excited into the hole of the K shell, finally an electron in L_3 shell is ejected as an Auger electron.

$$E_a = E_K - E_{L2} - E_{L3} - f(L_2, L_3) + R_{in} + R_{ex} - \Phi \quad (8.2)$$

E_K, E_{L2}, E_{L3}: binding energy of each energy level, $f(L_2, L_3)$: interaction energy between holes of x-orbital, R_{in}, R_{ex}: intra- and inter-atomic relaxation energy. Φ : work function

Fig. 8.1 Correlation in the binding energy between the corresponding levels in an Auger transition

kinetic energy of an Auger electron. It is basically determined by the difference in the binding energies of the three related electronic states. The fourth and subsequent terms are the correction terms, and this is the same idea as when calculating the energy of fluorescent X-rays in Sect. 4.2. Since the multiple electron holes contributes to the vacancy interaction energy of the fourth term to a certain extent, the kinetic energy of Auger electrons may derive from the energy difference between three energy levels of the relating electron orbitals. Also, since Auger electrons are emitted from a solid surface, the energy equivalent to the work function, Φ, explained in Sect. 4.2, is required. It is added at the end of the right side of Eq. (8.2).

8.2   Auger Electron Transition Process

Figure 8.2 illustrates changes in the electron configuration of the Mg atom during an Auger process. It starts from the place where an electron in the K shell is removed from the atom, which is induced by high-energy electron beam. See also the binding energy table of Table 4.1 in Chap. 4. Since the binding energy of the K-shell electron of Mg is 1305 eV, the incident electron beam needs to have more kinetic energy. When an electron hole is created in the K shell, an electron in the L shell is de-excited to fill it. At this time, the residual energy is emitted as a characteristic X-ray; in addition, another possibility is that the residual energy is used to emit an electron in the same

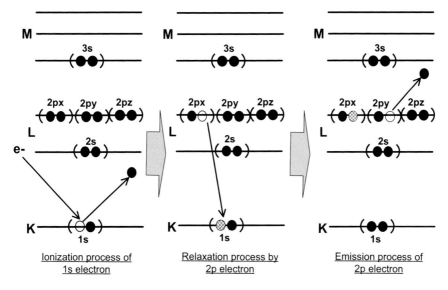

Fig. 8.2 Excitation/de-excitation paths of electrons in the KLL Auger transition of Mg atom

or different electron orbital to the outside of the system. Since the first relaxation process is exactly the same as the excitation mechanism of the characteristic X-ray, the electronic transition path should follow the selection rule for the transition of inner electron orbitals, explained in Sect. 5.2; however, Auger electrons are also observed through paths that do not follow this selection rule. On the other hand, the second relaxation process is not restricted by the selection rule, and any electron may be emitted if the energy balance is possible. In general, the first relaxation process that follows the selection rule may become a dominant path for Auger electron.

The notation of the Auger process is written in *the order of (electron energy level of the initial electron hole) (energy level that donates the filled electron) (energy level that emanates the Auger electron)*. Note that each of p, d, or f electron orbital creates two individual energy levels, as described in Sect. 5.1 (see Fig. 5.1). In the case of Mg, the process of filling a hole of the $1s_{1/2}$ energy level in the K shell with an electron of the $2p_{1/2}$ or $2p_{3/2}$ energy level in the L shell is allowed according to the selection rule and thus the Auger transition may be caused with high probability. There are several possibilities as to which electrons are eventually emitted as an Auger electron. When an electron of the $2p_{1/2}$ level in the L shell fills a hole of the $1s_{1/2}$ level in the K shell and then an electron in the same $2p_{1/2}$ level becomes emitted, the kinetic energy of the Auger electron is calculated to be $1305 - 52 - 52 = 1201$ eV. This transition is referred to as KL_2L_2. There are Auger electrons due to the KL_2L_3 and KL_3L_3 transitions, and these Auger electrons cannot be distinguished because the $L_2(2p_{1/2})$ and $L_3(2p_{3/2})$ levels of the Mg atom have the same binding energy (see Table 4.1). In addition, when an electron in the M shell ($3s_{1/2}$ level) fills an electron hole in the L shell ($2p_{1/2}$ or $2p_{3/2}$ level), an electron in the same $3s_{1/2}$ level may become an Auger electron having the kinetic energy: $52 - 2 - 2 = 48$ eV, which is the $L_{2,3}M_1M_1$ transition. In these calculations, the correction terms in Eq. (8.2) in Fig. 8.1 are ignored.

Chapter 9
Electron Spectroscopy for Materials Analysis

9.1 Photoelectron Spectrum

Various types of electrons are ejected from materials to be analyzed by irradiating high-energy electrons or X-ray. Generally, photoelectrons excited by the X-ray and Auger electrons excited by the electron beam as well as the X-ray are employed for materials analysis.

As already explained in Sect. 4.2, the kinetic energy of photoelectron directly relates to the binding energy of a certain energy level in a sample material via Eq. (4.2), although several correction terms in the equation should be taken into account. Therefore, the photoelectron spectrum is extensively measured for obtaining the analytical information, which is abbreviated to "XPS" (*X-ray photoelectron spectroscopy*). Let us investigate the XPS spectrum in more detail. Figure 9.1 shows an XPS spectrum of gold in a region of N_6 ($4f_{5/2}$) and N_7 ($4f_{7/2}$) energy levels, when monochomatic X-ray of Al Kα line (1486.6 eV) irradiates on a gold surface. Their binding energies are reported to be 88 and 84 eV [1]; therefore, the observed XPS peaks almost correspond to the binding energies, where the correction terms in Eq. (4.2) are ignored. In this case, a subtraction from the photon energy of Al Kα line has been carried out in the x-axis labeled as "Binding Energy". This spectrum of gold is usually used to calibrate the energy scale of the spectrometer as well as to estimate the resolution power. Measured kinetic energies of XPS are summarized for various elements and their energy levels in a data book [1].

Another example is an XPS spectrum of silicon in a wide range of the binding energy, as shown in Fig. 9.2. An as-received surface of a silicon wafer for semiconductor production is measured when irradiating the Al Kα X-ray on the sample. In addition of photoelectrons of Si $2s_{1/2}$ (148 eV) and $2p_{1/2, 3/2}$ (99 eV), O 1s (532 eV) and C 1s (284 eV), which are attributed to the oxide layer and surface contaminants, are dominantly observed in this spectrum. This measurement implies that the XPS spectrum is very sensitive to a state of the sample surface, which will be more explained in Sect. 9.3.

© The Author(s), under exclusive license to Springer Nature Singapore Pte Ltd. 2021
K. Wagatsuma, *Spectroscopy for Materials Analysis*, SpringerBriefs in Materials,
https://doi.org/10.1007/978-981-16-5946-1_9

Fig. 9.1 XPS spectrum of
gold in the 4f region

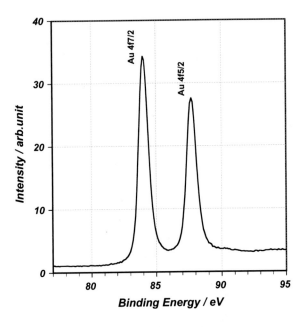

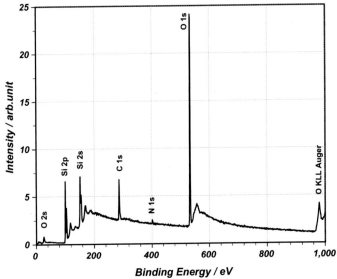

Fig. 9.2 XPS spectrum for an as-received surface of a silicon wafer

9.2 Auger Electron Spectrum

We will explain the Auger transition based on a result of the actual measurement in *Auger electron spectroscopy* (usually abbreviated to "AES"). Figure 9.3 shows an AES spectrum of nickel at kinetic energies of 600–900 eV. Auger transitions involving the L and M shells of nickel are observed in this region, and it can be seen that there are many AES peaks including small transition probabilities. This is due to a number of the Auger transition paths; among them, there are three major peaks with high intensity. These are assigned to $L_3M_{23}M_{23}$, $L_3M_{23}M_{45}$, and $L_3M_{45}M_{45}$. Their physical meaning and calculation from the binding energy are shown below. The work function is required for the calculation, for which the work function Φ of nickel metal adopts a literature value of 4.4 eV. In addition, three correction terms in Eq. (8.2) in Fig. 8.1 are collectively expressed as R. For the binding energies, use Table 4.1 in Sect. 4.1.

$L_3M_{23}M_{23}$: An electron hole is created in the $2p_{3/2}$ level in the L orbital, and then any electron in the $3p_{1/2}$ or $3p_{3/2}$ level in the M orbital fills the hole. Finally, other one electron in the $3p_{1/2}$ or $3p_{3/2}$ level is ejected as an Auger electron with the remaining energy. From the binding energy of each energy level, the kinetic energy of the electron is $E\ (L_3M_{23}M_{23}) = 855 - 68 - 68 + R - 4.4 = (714.6 + R)$ eV.

$L_3M_{23}M_{45}$: An electron hole is created in the $2p_{3/2}$ level in the L orbital, and then any electron in the $3p_{1/2}$ or $3p_{3/2}$ levels in the M orbital fills the hole. Finally, one electron in the $3d_{3/2}$ or $3d_{5/2}$ level is ejected as an Auger electron with the remaining energy. From the binding energy of each energy level, the kinetic energy of the electron is $E\ (L_3M_{23}M_{45}) = 855 - 68 - 4 + R - 4.4 = (778.6 + R)$ eV.

$L_3M_{45}M_{45}$: An electron hole is created in the $2p_{3/2}$ level in the L orbital, and then any electron in the $3d_{3/2}$ or $3d_{5/2}$ level in the M orbital fills the hole. Finally, other one electron in the $3d_{3/2}$ or $3d_{5/2}$ level is ejected as an Auger electron with the remaining energy. From the binding energy of each energy level, the kinetic energy of the electron is $E\ (L_3M_{45}M_{45}) = 855 - 4 - 4 + R - 4.4 = (842.6 + R)$ eV.

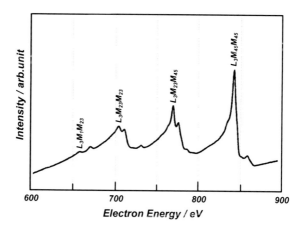

Fig. 9.3 AES spectrum of nickel after the surface is cleaned by argon ion bombardment

The measured kinetic energies for the Auger transitions of $L_3M_{23}M_{23}$, $L_3M_{23}M_{45}$, and $L_3M_{45}M_{45}$ have been reported. Although there are slight differences depending on the literatures, E $(L_3M_{23}M_{23})$ = 704.5 eV, E $(L_3M_{23}M_{45})$ = 776.7 eV, E $(L_3M_{45}M_{45})$ = 841.8 eV are published as one literature value. The difference between the measured and the above calculated values is considered to be the contribution of the correction terms. The correction terms are also dependent on each Auger transition and contribute to the kinetic energy more largely compared to XPS, because the Auger process involves two electron holes having different interaction energies. The measured values are compiled in data books published by manufacturers of AES equipment [2, 3].

In addition, many channels for the Auger transition are found in Fig. 9.3. Those relating to the M_1 ($3s_{1/2}$) level appear on the low energy side, and those involving the L_2 ($2p_{1/2}$) level appear on the slightly higher energy side than the L_3 ($2p_{3/2}$) level. In addition, there may appear Auger transitions relating to the valence band of nickel metal. The outermost electrons of nickel occupy the N_1 ($4s_{1/2}$) level, but this level would exist exactly in the case of a free atom (gaseous state), and in the metallic state, this orbital is involved in a valence band, where free electrons are co-shared as the whole. The binding energy table shown as Table 4.1 in Sect. 4.1 does not show the binding energy of the electrons in the valence band; in the case of nickel, it is about the same as that of the M_{45} level.

Figure 9.4 shows a measured AES spectrum of metallic silicon in an energy range of 0–120 eV. Let us consider how they are interpreted in the same way as the nickel AES spectrum of Fig. 9.3. The AES spectrum of silicon is attributed to the KLL Auger transition, where an electron in the L shell fills the electron hole in the K shell through several different channels. We can estimate the kinetic energies of each peak from the binding energies of the energy levels, which are found in Table 4.1.

Fig. 9.4 AES spectrum of silicon after the surface is cleaned by argon ion bombardment

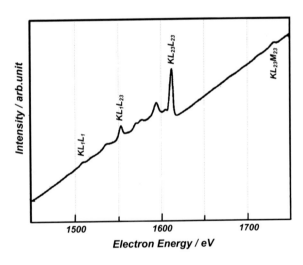

References

1. Wagner CD (1989) NIST X-ray photoelectron spectroscopy database. Office of Standard Reference Data, NIST, Gaithersburg
2. Davis LE et al (1976) Handbook of Auger electron spectroscopy. Perkin Elmer
3. Sekine T et al (1982) Handbook of Auger electron spectroscopy. JEOL

Chapter 10
Application of Electron Spectroscopy

10.1 Surface Analysis and Local Analysis

The greatest feature of AES is derived from the use of electrons as both the excitation and the detection tool. As mentioned in Sect. 7.2, the electron beam decays rapidly due to its large interaction with the substance, and the measuring area in which it is involved does not extend so widely. This behavior becomes noticeable near the sample surface on which an electron beam is irradiated. See Fig. 10.1. This figure schematically indicates how deep an electron with a kinetic energy of 3 keV affects when it enters a silicon surface as a trajectory. The electron is blocked on the sample surface, but it gives the kinetic energy to the colliding atoms, resulting in their ionization through cascade collisions. It shows that the range of influence is about 50 nm from the surface; of course, the range may be dependent on the kinetic energy of electron, the kind of the sample, and the state of the surface. However, it is more important that most of the generated electrons disappear in the sample and then cannot escape from the surface, and that the kinetic energy is reduced in the deep portion to the extent that the Auger process can hardly occur. The depth at which the generated Auger electrons can jump out into a vacuum is estimated by an *escape depth*, which is determined by the kinetic energy of Auger electrons. Figure 10.2 indicates that the escape depth ranges from 0.5 to a few nm [1], implying that *the AES information is restricted in several atomic layers from the surface.* This is inconvenient to obtain an overall picture of the sample because it provides only information near the surface; conversely, it is extremely useful for obtaining analytical information specialized for the surface. This is called "Surface Analysis". Compared to AES, incident X-ray in XPS can penetrate into a sample surface on the order of several μm depth due to its weaker interaction with the substance. However, the escape depth of photoelectron is almost similar to that of Auger electron, because their kinetic energies are in a range of a few keV or less. Therefore, *AES and XPS are both categorized into surface analysis.*

Figure 10.3 shows an XPS spectrum of a silicon wafer, which is taken with the same sample and the same measuring conditions as the spectrum of Fig. 9.2 but

© The Author(s), under exclusive license to Springer Nature Singapore Pte Ltd. 2021
K. Wagatsuma, *Spectroscopy for Materials Analysis*, SpringerBriefs in Materials,
https://doi.org/10.1007/978-981-16-5946-1_10

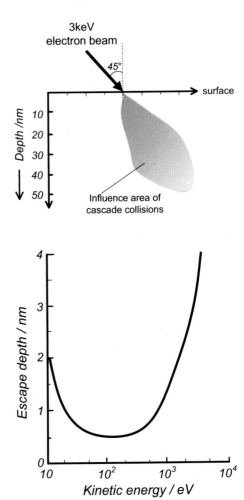

Fig. 10.1 Schematic presentation of the influence zone when injecting an electron of 3 keV at an incident angle of 45°

Fig. 10.2 Variation in the escape depth of electrons emitted from solid surfaces as a function of the kinetic energy

the topmost surface is removed by argon ion bombardment, called *sputtering*. The sputtering by argon ions, where the high-energy ion beam irradiates on the sample surface in vacuum atmosphere, is a common technique combined with XPS and AES, which leads to etching of the sample surface at a rate of a few nm/min. It can be seen that O 1s and C 1s peaks almost disappear whereas the intensity of the Si signals becomes larger in this spectrum, clearly because the surface contaminants are cleaned by the argon ion etching. In this case, the removed layer has a thickness of several nm, indicating that XPS provides the elemental information on only the sample surface. In fact, such surface information often plays a core role in material analysis. For example, both XPS and AES have made many contributions to the analysis of industrial materials such as passivation thin films of steel and semiconductor film formation.

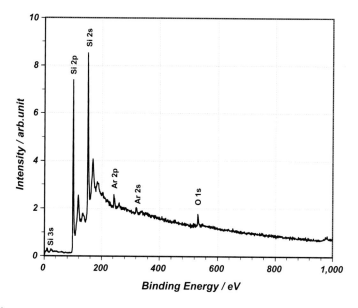

Fig. 10.3 XPS spectrum of a silicon wafer whose surface is etched by argon ion bombardment

Especially in XPS, the kinetic energy of photoelectron includes another piece of information for material analysis. Figure 10.4 shows an XPS spectrum of

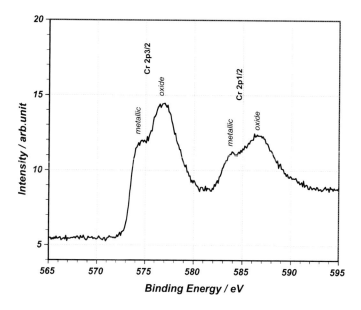

Fig. 10.4 XPS spectrum of Cr $2p_{1/2}$ and $2p_{3/2}$ peaks in SUS304 stainless steel

Fig. 10.5 XPS depth profile of an oxide layer on the surface of Cr–Co–Mo alloy. The sample is treated in a glow discharge plasma with 80-Pa argon and 10-Pa oxygen at a discharge current of 70 mA for 6 min. The sputtering rate of argon sputtering is about 1.5 nm/min. This article is reprinted from [3] under the permission of Elsevier

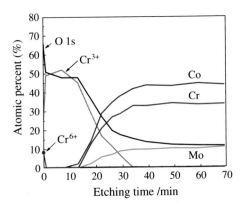

chromium, Cr $2p_{1/2}$ and Cr $2p_{3/2}$ peaks, for an as-received surface of SUS304 stainless steel, whose chemical composition is 18% Cr, 8% Ni, and 73% Fe. These peaks of chromium appear dominantly rather than the corresponding XPS peaks of iron, which is the evidence that Cr is enriched on the surface due to a formation of Cr_2O_3 surface layer. Furthermore, as shown in Fig. 10.4, the XPS spectrum comprises two peaks each for the Cr $2p_{1/2}$ and Cr $2p_{3/2}$. The reason for this splitting is that their kinetic energies are somewhat different for the chemical state of chromium, between metal and oxide. For instance, the peak positions are 574 eV for the metallic state and 577 eV for the oxide in the Cr $2p_{3/2}$ peak. This effect is called "Chemical Shift". The degree of the chemical shift in XPS changes for the kind of compounds of a certain element, and thus provides useful information when the chemical state will be determined. The chemical shift has been investigated in detail for many elements and their compounds [2].

Changes in the intensity of XPS or AES during ion sputtering represent the in-depth distribution of elements and their chemical species in a sample, which is called a depth profile. This becomes a powerful tool in the surface analysis of various materials. Figure 10.5 shows a typical example of the depth profile in XPS using argon ion sputtering [3]. The sample is an oxide layer, which is formed in a glow discharge plasma with argon-oxygen mixed gas, on a chromium–cobalt–molybdenum alloy. This type of alloy is now employed as one of metallic biomaterials due to its high strength and good corrosion resistance. The depth profile indicates that the oxide layer comprises a topmost portion of CrO_3 and a major portion of Cr_2O_3 on the alloy surface, which are distinguished from their chemical shifts in the Cr $2p_{3/2}$ peak. It indicates that the chromium oxide is preferentially formed on the surface and contributes to the corrosion resistance in this alloy.

Another feature of AES is the control of the incident electron beam. Since an electron beam is a flow of electrons with a negative charge, the irradiated area can be narrowed down by applying an electric field and/or a magnetic field. Recent devices have achieved a beam diameter of several hundred nm or less. In materials science, the analytical investigation may be important for nm-sized targets such as fine precipitates and grain boundaries. AES is successfully applied for this purpose, which is called "Local Analysis".

10.2 Equipment for Electron Spectroscopy

The device is an electron source making an electron beam or a primary X-ray source, an energy analyzer that separates emitted electrons by kinetic energy, and an electron detector that converts the number of electrons into an electronic signal. Figure 10.6 illustrates a block diagram of a spectrometer equipped with a *hemispherical-type analyzer* that has been widely employed for the measurement in XPS. This system obtains monoenergetic electrons by applying an electric field between two bowl-like electrodes and by focusing them on the exit slit, which realizes the energy dispersion of electrons. It provides the XPS signal with high sensitivity and spectral resolution. Figure 10.7 shows a schematic drawing of a *cylindrical mirror analyzer* (CMA), which is a typical energy analyzer in AES. The trajectory of electrons can be changed by changing an electric field applied between the inner and outer cylinders of the analyzer. The electric field makes it possible to pass only electrons with a specific kinetic energy to the detector, since the focus position is a function of the kinetic energy of electrons. The central part of the hollow cylindrical analyzer is empty in this system, so the device is made compact by incorporating an electron source for AES in this portion. Several types of detectors are employed to amplify the number of electrons up to a level that it can be detected as a current in the electron spectrometer, such as an electron multiplier that consists only of the dynode electrodes of the scintillation tube for X-ray fluorescence spectroscopy (see Fig. 6.7).

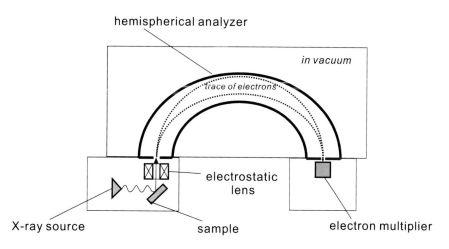

Fig. 10.6 XPS spectrometer system having a hemispherical-type energy analyzer

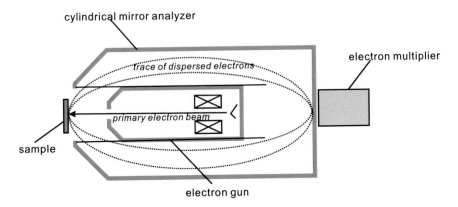

Fig. 10.7 AES spectrometer system having a cylindrical mirror energy analyzer

References

1. Someno D, Yasomori I (1976) Surface analysis. Kodansha Scientific, Tokyo
2. Briggs D, Seah MP (1990) Practical surface analysis, vol 1. Wiley
3. Furukawa K, Wagatsuma K (2020) Surf Interfaces 18:100402

Chapter 11
Electronic Transition Between Outer Shell Electron Orbitals

11.1 Excitation and De-excitation

In both X-ray fluorescence spectroscopy (Chaps. 4–6) and Auger and photo electron spectroscopy (Chaps. 8–10), the electron ionization (electron vacancy) in an inner shell orbital is first caused by introducing X-ray and electron beam. Because the electron in the inner shell is strongly constrained by the nucleus, it is necessary to provide an appropriate high energy for the ionization. Its energy is on the order of 1000 eV and can be obtained by X-ray or accelerated electron beams. Then, is there any spectrometric method other than these spectroscopies that contributes to the qualitative and quantitative analysis? The answer is that there is another powerful spectroscopy, as already explained a bit in Sect. 2.3.

Figure 11.1 shows the electron configuration of the Mg atom. The outermost electron is stored in the 3s orbital in the ground state; however, by supplying a little energy, this electron can be raised to the vacant 3p, 3d orbitals or the N-shell orbitals. The energy required for these phenomena is a few eV–10 eV, which is much smaller than that of the inner shell, and this energy can be supplied by an external source such as an electric discharge without using a high-energy source like an electron beam. Since the raised electron to the 3p orbital is very unstable, it quickly returns to the 3s orbital. At this time, the radiation corresponding to the energy difference between these electron orbitals is emitted. This process is *excitation/de-excitation* in a narrow sense and can be expressed by Eq. (11.1) shown in Fig. 11.1. This spectrum normally appears in a wavelength range of 100–1000 nm due to a small energy difference between the electron orbitals involved. The wavelength range is called the *ultraviolet/visible radiation*, but it is more common to use "atomic emission/absorption spectroscopy" rather than ultraviolet/visible spectroscopy.

K. Wagatsuma, *Spectroscopy for Materials Analysis*, SpringerBriefs in Materials, https://doi.org/10.1007/978-981-16-5946-1_11

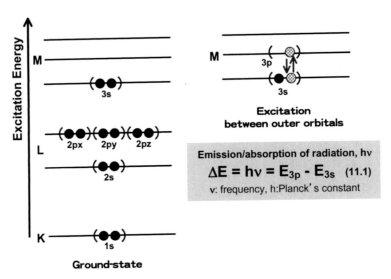

Fig. 11.1 Electron transition for a spectral line in the electron configuration of magnesium atom

11.2   Atomic Emission Spectrum of Hydrogen Atom

Using an example of a hydrogen atom, we explain the atomic emission spectrum in more detail. An electron in which a certain transition is involved is in the outermost orbital, whereas the inner shell electrons do not move at all and thus they can be ignored in the electron transition. However, since the hydrogen atom has only one electron, there is no distinction between the inner and the outer shells. The most stable electron configuration of a hydrogen atom is 1s, which is the ground state electron configuration. If this electron moves to any one outer orbital, 2s, 2p, and so on, all of them become the excited electron configurations, and when an electron transition occurs between these electron configurations, the energy difference is observed as an electromagnetic wave. Figure 11.2 illustrates paths of the electronic transitions in the ground/excited energy levels of the hydrogen atom, called an *energy level diagram* [1]. Table 11.1 shows a table of the electron configuration of the hydrogen atom [2], called an *energy level table*. It is important in this diagram how the standard of energy should be set. Considering that an electron orbital has a specific energy corresponding to the potential energy with respect to the atomic nucleus, the most stable state of the 1s orbital has the largest (deepest) energy for the electron to be released from the constraint of the atom. The energy axis in Fig. 11.2 is determined according to this idea. The level labeled as 1, which appears at the bottom of this figure, is derived from the 1s orbital with an energy value of roughly -110×10^3 cm^{-1}. On the other hand, when considering an electronic transition, the energy difference between the relating energy levels always determines the appearance of the spectral line (the wavelength);

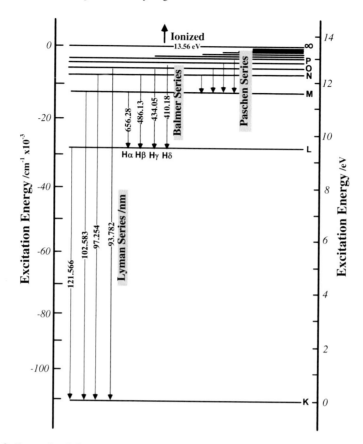

Fig. 11.2 Energy level diagram of hydrogen atom regarding the electron orbitals

therefore, it does not matter where the reference of the energy axis is taken. According to this idea, the energy of the 1s orbital is set to zero and then the energy of other outer electron orbitals is represented on the positive scale of energy, as shown in Table 11.1. This scale is generally used to describe the assignment of spectral lines. The spectrum of the hydrogen atom is attributed to any electronic transition between the energy levels summarized in Table 11.1 [2]. However, the initial and final states for an electronic transition are not the electron orbitals of the hydrogen atom themselves, but the energy levels derived from the electron orbitals. Also, all of the transition paths between these levels are not allowed and some are forbidden. The *spectral term splitting* and the *selection rule*, which are already described in Sects. 5.1 and 5.2, should be taken into consideration in more detail. In Chap. 12, we will deal with these important principles in the atomic spectrum, but let us calculate wavelengths of the spectral lines from the energy level table in the case of the hydrogen atom, even at some inaccuracies.

Table 11.1 Energy levels for neutral hydrogen atom

Config	Term	Level (cm^{-1})
1s	1s $^2S_{1/2}$	0.000
2s	2s $^2S_{1/2}$	82,258.942
2p	2p $^2P_{1/2}$	82,258.907
	2p $^2P_{3/2}$	82,259.272
3s	3s $^2S_{1/2}$	97,492.208
3p	3p $^2P_{1/2}$	97,492.198
	3p $^2P_{3/2}$	97,492.306
3d	3d $^2D_{3/2}$	97,492.306
	3d $^2D_{5/2}$	97,492.343
4s	4s $^2S_{1/2}$	102,823.839
4p	4p $^2P_{1/2}$	102,823.835
	4p $^2P_{3/2}$	102,823.881
4d	4d $^2D_{3/2}$	102,823.881
	4d $^2D_{5/2}$	102,823.896
4f	4f $^2F_{5/2}$	102,823.896
	4f $^2F_{7/2}$	102,823.904
5s	5s $^2S_{1/2}$	105,291.617
5p	5p $^2P_{1/2}$	105,291.615
	5p $^2P_{3/2}$	105,291.638
5d	5d $^2D_{3/2}$	105,291.638
	5d $^2D_{5/2}$	105,291.646
5f	5f $^2F_{5/2}$	105,291.646
	5f $^2F_{7/2}$	105,291.650
5g	5g $^2G_{7/2}$	105,291.650
	5g $^2G_{9/2}$	105,291.652
6s	6s $^2S_{1/2}$	106,632.136
6p	6p $^2P_{1/2}$	106,632.135
	6p $^2P_{3/2}$	106,632.148
6d	6d $^2D_{3/2}$	106,632.148
	6d $^2D_{5/2}$	106,632.152
6f	6f $^2F_{5/2}$	106,632.152
	6f $^2F_{7/2}$	106,632.155
6g	6g $^2G_{7/2}$	106,632.155

(continued)

Table 11.1 (continued)

Config	Term	Level (cm^{-1})
	6g ^{2}G$_{9/2}$	106,632.156
6h	6h ^{2}H$_{9/2}$	106,632.156
	6h ^{2}H$_{11/2}$	106,632.157
7s	7s ^{2}S$_{1/2}$	107,440.425
To	To	To
7i	7i ^{2}I$_{13/2}$	107,440.439
8s	8s ^{2}S$_{1/2}$	107,965.036
To	To	To
8j	8j ^{2}J$_{15/2}$	107,965.045
9s	9s ^{2}S$_{1/2}$	108,324.706
To	To	To
9k	9k ^{2}K$_{17/2}$	108,324.708

The ground state is 1s^{2}S$_{1/2}$. The numerical values are extracted from a data book [1]

Figure 11.2 includes a series of spectral lines assigned to the electronic transitions from/to the ground state level 1 (1s orbital). This is called "the Lyman series" [1]. Let us calculate these wavelengths using the energy level table in Table 11.1 [2]. In Table 11.1, ignore the *term* of the column 2 and focus on the energy values in the column 3. The unit is cm^{-1}. The *Lyman series* spectrum results from transitions from the p-orbitals of the outer electron shells L, M, N, and so on, to the s orbital of the K shell. When these spectral lines are calculated by reading each energy value in Table 11.1, their wavelengths are as follows:

2p → 1s: $1/(82,258.9 - 0) = 121.57 \times 10^{-7}$ cm $= 121.57$ nm,

3p → 1s: $1/(97,492.2 - 0) = 102.57 \times 10^{-7}$ cm $= 102.57$ nm,

4p → 1s: $1/(102,823.84 - 0) = 97.25 \times 10^{-7}$ cm $= 97.25$ nm.

These wavelength values agree to the observed values in an actual spectrum of hydrogen atom, as indicated in Fig. 11.3. Similarly, the *Balmer series* and the *Paschen series* found in Fig. 11.2 are assigned to electronic transitions in which the 2s and 3s orbitals are involved [1], respectively.

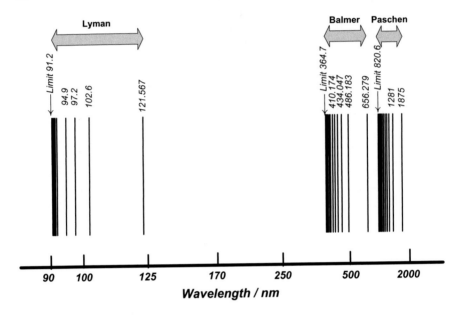

Fig. 11.3 Spectrograph of emission lines of hydrogen atom

References

1. Herzberg G (1944) Atomic spectra and atomic structure. Dover Publications, New York
2. Moore CE (1949) Atomic energy levels, vol 1. NBS Circular No. 467. U.S. Government Printing Office, Washington, DC

Chapter 12
Principle of Atomic Emission Spectroscopy

12.1 Spectral Term Splitting

In Sect. 11.2, we state that the initial and final states of a certain electronic transition that appears as the atomic spectrum are not the electron orbitals themselves but the energy levels derived from them. In general, one electron orbital may produce several energy levels, called "spectral term splitting". The reason for this is that the interaction between the azimuthal momentum and the spin angular momentum of each orbital creates multiple energy states. The same explanation was given on the X-ray levels in Sect. 5.1; in the case of the inner shell ionization, only the orbital where an electron hole occurs is considered. However, in outermost shell orbitals, the situation is more complicated because other orbitals around the orbital relating to a particular transition may affect the spectral term splitting. Moreover, such splitting should be taken into consideration not only for the initial state but also for the final state of the electronic transition. We will see the spectral term splitting using a typical example.

In the concept of quantum chemistry, it is possible to characterize each electron orbital with a combination of four quantum numbers. Recall the explanation for the quantum numbers summarized in Fig. 2.5 in Sect. 2.2. The shape of the orbital and the total momentum of electron are determined by the quantum numbers. When an electron is filled in an individual orbital, its azimuthal angular momentum and spin angular momentum interact with those of other electrons in other orbitals to generate multiple energy levels. Strictly speaking, it is necessary to consider all the electrons in the atom system and then evaluate the energy states; however, this task is very difficult and actually impossible. Therefore, there are several approximate treatments presented, which can explain the term splitting for energy levels that are actually observed, if applicable elements is considered adequately. "Russell–Saunders coupling", also known as the "LS coupling", is a good approximation for light elements with atomic numbers up to about 40 with the exception of inert gas atoms.

© The Author(s), under exclusive license to Springer Nature Singapore Pte Ltd. 2021
K. Wagatsuma, *Spectroscopy for Materials Analysis*, SpringerBriefs in Materials,
https://doi.org/10.1007/978-981-16-5946-1_12

Russell-Saunders coupling

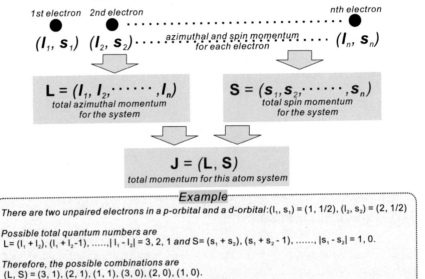

Fig. 12.1 Spectral term splitting based on the Russell–Saunders coupling (LS coupling)

Figure 12.1 indicates the concept of the Russell–Saunders coupling. Azimuthal angular momentum (*azimuthal quantum number l*) and spin angular momentum (*spin quantum number s*) are the physical quantities attached to individual electrons. In this approximation, the azimuthal angular momentum and the spin angular momentum are each extracted from all electrons in an atom system, to obtain the total angular momentum. First, the azimuthal and the spin momentums are combined separately. For each coupling of the angular momentum, the *total azimuthal quantum number L* and the *total spin quantum number S* are obtained according to the concept of *energy quantization*. Finally, the total azimuthal and spin quantum numbers are combined to obtain the *quantum number J of the entire system*. Since this quantum number determines the energy of the atom system, it represents an energy level. A concrete example may help us understand the Russell–Saunders coupling, as described below.

When considering the electron configuration of an atom, it is an important fact that any electron orbitals filled with two electrons having a spin pair of $+1/2$ and $-1/2$ (*closed sub-shell*) create no angular momentum in the atom system, because all the angular momentum cancel each other in the sub-shell. That is, they can be ignored in the estimation of the LS coupling, and also electrons in the inner shells need not be considered due to their spin pairs. Only electrons that do not form a spin pair in the outermost shell, which are called *unpaired electrons*, must be considered. As an example, if there are two unpaired electrons, one in the d-orbital and one in the l-orbital, how is the spectral term splitting evaluated based on the LS coupling? The column inserted in Fig. 12.1 shows a procedure for finding the spectral terms and the energy levels in this system. First, since each combination of (l, s) is that the d-orbital is $(2, 1/2)$ and the p-orbital is $(1, 1/2)$, the total azimuthal quantum number L and the total spin quantum number S are obtained by combining each quantum number under the quantizing condition, resulting in $L = 3, 2, 1$ and $S = 1, 0$. Further, the quantum number J of the entire system is calculated for each combination between L and S. For instance, in the case of $(L, S) = (3, 1)$, there are three types of energy levels, $J = 4, 3, 2$. Similarly, all the combinations result in 12 different energy levels.

12.2 Notation of Spectral Terms in LS Coupling

It is summarized in Fig. 12.2. Since this is idiomatically determined, we simply follow the rule. For example, an energy level of $(L, S, J) = (3, 1, 4)$ is described as $^3F_4$, and an energy level of $(2, 0, 2)$ as $^1D_2$. First, the value of L is written using the corresponding uppercase alphabet, and a value of *spin multiplicity* $(2S + 1)$ in the left superscript position, and the value of J in the right subscript position.

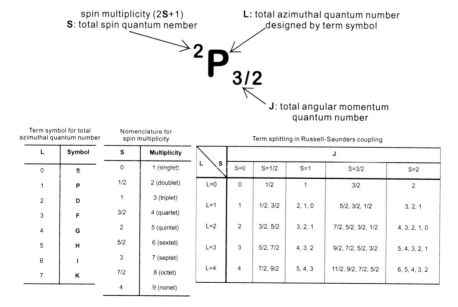

Fig. 12.2 Notation of the spectral terms in the Russell–Saunders coupling

12.3   Selection Rule for LS Coupling

Up to the previous sections, we have explained all the items to understand the energy level. As already mentioned repeatedly, the energy difference between the upper and lower energy levels for an electronic transition is converted into the wavelength of radiation, and then it provides us with analytical information. If there were no restrictions on the electronic transitions, a large number of the combinations could appear between the energy levels; however, there are several restrictions on the combination of energy levels that allows electronic transitions. This is called the "Selection Rule" [1]. Figure 12.3 summarizes the selection rule for electronic transitions between outer shell electron orbitals, when a variation of the electric dipole causes the radiation. It consists of six items, and a transition path that satisfies all the rules is described as an *allowed transition* and the others as a *forbidden transition*. The selection rule (3) is related to the magnetic quantum number, which is not dealt with in the scope of this book. The selection rules (1) and (2) strictly apply to all electronic transitions; on the other hand, the rules (4)–(6) are fully effective if the LS coupling is strictly

Selection rule for electric dipole radiation

(1) **Parity forbidden**
The symmetry of the electron orbital must be different before and after an electron transition.

(Ex) s-orbital $\leftrightarrow$ s-orbital : forbidden, s-orbital $\leftrightarrow$ p-orbital : allowed, s-orbital $\leftrightarrow$ d-orbital : forbidden, p-orbital $\leftrightarrow$ d-orbital : allowed, p-orbital $\leftrightarrow$ f-orbital : forbidden

(2) **Total quantum number (J)**
Change in the total quantum number must be -1, 0, or +1 in an electron transition. $\Delta J = \pm 1, 0$
A transition from 0 to 0 is forbidden. $J = 0 \not\leftrightarrow J = 0$.

(Ex) $J=0 \leftrightarrow J=1$: allowed, $J=1 \leftrightarrow J=2$: allowed, $J=3 \leftrightarrow J=2$: allowed, $J=0 \leftrightarrow J=0$: forbidden, $J=0 \leftrightarrow J=2$: forbidden, $J=1 \leftrightarrow J=3$: forbidden, $J=1/2 \leftrightarrow J=5/2$: forbidden, $J=3/2 \leftrightarrow J=7/2$: forbidden

(3) **Magnetic quantum number (m)**
Change in the magnetic quantum number must be -1, 0, or +1 in an electron transition.

When Russell-Saunders coupling is strictly satisfied,
(4) **Total azimuthal quantum number (L)**
Change in the total azimuthal quantum number must be -1, 0, or +1 in an electron transition. $\Delta L = \pm 1, 0$
A transition from 0 to 0 is forbidden. $L = 0 \not\leftrightarrow L = 0$

(Ex) $L=0 \leftrightarrow L=1$: allowed, $L=1 \leftrightarrow L=2$: allowed, $L=3 \leftrightarrow L=2$: allowed, $L=0 \leftrightarrow L=0$: forbidden, $L=0 \leftrightarrow L=2$: forbidden, $L=1 \leftrightarrow L=3$: forbidden, $L=2 \leftrightarrow L=4$: forbidden

(5) **Total spin quantum number (S)**
Change in the total spin quantum number must be -1 or +1 in an electron transition. $\Delta S = 0$

(Ex) $S=0 \leftrightarrow S=0$: allowed, $S=1/2 \leftrightarrow S=1/2$: allowed, $S=1 \leftrightarrow S=1$: allowed, $S=1 \leftrightarrow S=2$: forbidden, $S=1/2 \leftrightarrow S=3/2$: forbidden, $S=3/2 \leftrightarrow S=5/2$: forbidden, $S=3/2 \leftrightarrow S=7/2$: forbidden

(6) **Sub-shell of electron orbital (*l*)**
Change in the azimuthal quantum number must be -1, or +1 in an electron transition. $\Delta l = \pm 1$

(Ex) s-orbital $\leftrightarrow$ p-orbital: allowed, p-orbital $\leftrightarrow$ d-orbital : allowed, d-orbital $\leftrightarrow$ f-orbital : allowed, s-orbital $\leftrightarrow$ f-orbital : forbidden, p-orbital $\leftrightarrow$ g-orbital: forbidden

Fig. 12.3 Selection rule for the dipole transition between energy levels under the condition of the Russell–Saunders coupling

held; otherwise, forbidden spectral lines may be observed (when it breaks out of this approximation). Because these are itemized rules, it is not difficult to apply them and find a spectral line that follows the selection rule. However, it is difficult to explain the physical meaning of the selection rule intuitively. Here, the symmetry of the wave function, which relates to the selection rules (1) and (6), is explained briefly.

Whether an electron can move through a certain transition path is a matter of probability and can be evaluated by the *transition probability* [2]. Quantum chemistry shows that the transition probability is determined by what the product of the wave function in the initial state, the dipole moment function of the electron orbital, and the wave function in the final state is integrated over the entire space. In this calculation, the symmetry of the electron orbitals is an important factor. For example, if we integrate a function of $y = x$ from -1 to $+1$, the solution is zero. The answer is the same even if the integration interval is extended from $-\infty$ to $+\infty$. If $y = x^2$, the solution is not zero. In this way, whether the transition probability is zero or not can be easily distinguished from the symmetry of the function (*even function* or *odd function*). What is the symmetry of the electron orbitals? We can find the symmetry by the sign of the wave function being unchanged or changed at the position of the nucleus, and also it can be easily judged from the shape of the orbital. That is, the s orbital is even, the p orbital is odd, the d orbital is even, and the f orbital is odd, and the selection rule is judged from their symmetry. In Fig. 12.3, the selection rule (1), known as *parity forbidden*, describes the symmetry of all eigenfunctions of the electron orbitals. Note that the dipole moment function has a character of odd, and therefore, this rule can be understood within a mathematical common sense; that is, (even) × (odd) × (even) = (odd), (odd) × (odd) × (odd) = (odd), and (even) × (odd) × (odd) = (even), where the symmetry of the initial state, the dipole moment, and the final state is described. Unless the symmetry of the wave function changes before and after the electronic transition, the transition probability is always integrated to be zero. Further, the rule (6) is related to the symmetry of each orbital for the initial and final states in an electronic transition. The selection rule will be explained using an example of electron transitions in detail in Chap. 13.

References

1. Herzberg G (1944) Atomic spectra and atomic structure. Dover Publications, New York
2. Corney A (1977) Atomic and laser spectroscopy. Clarendon Press, Oxford

Chapter 13
Spectrum in Atomic Emission Spectroscopy

13.1 Atomic Spectrum of Aluminum

Figure 13.1 shows the electron configuration of the aluminum atom. Let us consider the atomic spectrum based on the electron arrangement. The ground state configuration is $(1s)^2(2s)^2(2p)^6(3s)^2(3p)$ with 13 electrons. The orbitals in which electrons are paired and filled in the inner shells are not involved in the spectral term splitting; therefore, only one electron in the outermost shell orbital of 3p should be considered. In this case, the electron has $(l, s) = (1, 1/2)$, directly leading to the total azimuthal and spin quantum numbers. That is, $(L, S) = (1, 1/2)$. Finally, the quantum number for the total angular momentum, J, results in two energy levels, such as $J = (L + S)$, $|L − S| = 3/2, 1/2$. This is the same idea as the X-ray level explained in Sect. 5.1. When these levels are written in the LS coupling notation, they become $^2P_{3/2}$ and $^2P_{1/2}$. Table 13.1 shows an energy level table of the Al atom [1]. The term representation and observed energy values of the energy levels are summarized. It is found in Table 13.1 that the energy of the corresponding levels is 112.04 cm^{-1} for the $^2P_{3/2}$ and 0.00 cm^{-1} for the $^2P_{1/2}$.

The electron configuration just above the ground state is referred to as the first excited state or the first excited level. In the case of the Al atom, the excitation energy of the 4s orbital is lower than that of the 3d orbital, and thus it is the first excited electron configuration. Since the spectral term of an electron in the 4s orbital is $(l, s) = (0, 1/2)$ and then $(L, S) = (0, 1/2)$ in the LS coupling, the quantum number of the whole system is $J = 1/2$ alone. Therefore, it forms only one energy level of $^2S_{1/2}$, which corresponds to an energy level of 25,347.69 cm^{-1} among the energy levels of Table 13.1. In Table 13.1, we can see that energy levels derived from the $(3s)(3p)^2$ electron configuration is located next to the $(3s)^2(4s)$ configuration. This is an electron configuration with three unpaired electrons, and it may be split into several energy levels, of which $^4P_{1/2,3/2,5/2}$ occupy the second excited energy position.

K. Wagatsuma, *Spectroscopy for Materials Analysis*, SpringerBriefs in Materials, https://doi.org/10.1007/978-981-16-5946-1_13

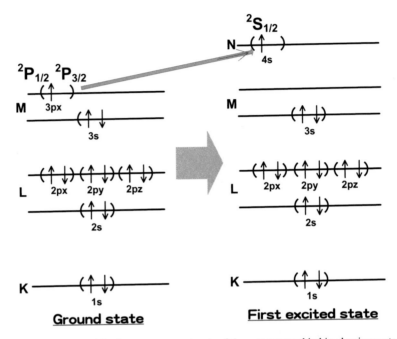

Fig. 13.1 Electron transition between energy levels of the outermost orbital in aluminum atom

Figure 13.2 shows the wavelength of the measured atomic spectrum of aluminum. Which energy levels are involved in the electronic transitions for these spectral lines? In this estimation, we will take into account all the selection rules except (3) in Fig. 12.3, assuming that the LS coupling is strictly followed. The first important condition is the symmetry of electron orbitals in the initial and final states of a certain electronic transition. It is possible to determine whether an electronic transition is allowed or forbidden under the selection rules (1) and (6). In the case of the aluminum atom, which energy levels in the excited electron configurations may couple with the two energy levels, $^2P_{3/2}$, $^2P_{1/2}$, of the ground-state electron configuration $(3s)^2(3p)$? An electron in the first excited electron configuration, $(3s)^2(4s)$, can be transited to the ground state, such as $(4s) \rightarrow (3p)$, because this event occurs from an even state to an odd state. Also, the electronic transition from the s orbital to the p orbital follows the selection rule (6). Further, changes in three quantum numbers should be taken into consideration. That is, the proposition is whether $(4s)\ ^2S_{1/2} \rightarrow (3p)\ ^2P_{1/2}$ or $(4s)\ ^2S_{1/2} \rightarrow (3p)\ ^2P_{3/2}$ is possible. The change in the quantum numbers is $\Delta J = 0$, $\Delta L = 1$, $\Delta S = 0$ in the former, and $\Delta J = 1$, $\Delta L = 1$, $\Delta S = 0$ in the latter; thus, all of them satisfy the selection rule shown in Fig. 12.3. Therefore, it is expected that aluminum spectral lines will appear from these transitions, as referred to their energy values in Table 13.1:

Table 13.1 Energy levels for neutral aluminum atom

Config	Term	Level (cm^{-1})
$3s^2 3p$	$3p \ ^2\mathbf{P}_{1/2}$	0.000
	$3p \ ^2\mathbf{P}_{3/2}$	112.04
$3s^2 4s$	$4s \ ^2\mathbf{S}_{1/2}$	25,347.69
$3s3p^2$	$3p^2 \ ^4\mathbf{P}_{1/2}$	29,020.22
	$3p^2 \ ^4\mathbf{P}_{3/2}$	29,066.90
	$3p^2 \ ^4\mathbf{P}_{5/2}$	29,142.68
$3s^2 3d$	$3d \ ^2\mathbf{D}_{3/2}$	32,435.45
	$3d \ ^2\mathbf{D}_{5/2}$	32,436.79
$3s^2 4p$	$4p \ ^2\mathbf{P}_{1/2}$	32,949.84
	$4p \ ^2\mathbf{P}_{3/2}$	32,965.67
$3s^2 5s$	$5s \ ^2\mathbf{S}_{1/2}$	37,689.32
$3s^2 4d$	$4d \ ^2\mathbf{D}_{3/2}$	38,929.42
	$4d \ ^2\mathbf{D}_{5/2}$	38,933.96
$3s^2 5p$	$5p \ ^2\mathbf{P}_{1/2}$	40,271.98
	$5p \ ^2\mathbf{P}_{3/2}$	40,277.92
$3s^2 4f$	$4f \ ^2\mathbf{F}_{5/2}$	41,318.74
	$4f \ ^2\mathbf{F}_{7/2}$	41,318.74
$3s^2 6s$	$6s \ ^2\mathbf{S}_{1/2}$	42,144.84
$3s^2 5d$	$5d \ ^2\mathbf{D}_{3/2}$	42,233.72
	$5d \ ^2\mathbf{D}_{8/2}$	42,237.71
$3s^2 6p$	$6p \ ^2\mathbf{P}_{1/2}$	43,334.95
	$6p \ ^2\mathbf{P}_{3/2}$	43,337.77
$3s^2 5f$	$5f \ ^2\mathbf{F}_{5/2}$	43,831.08
	$5f \ ^2\mathbf{F}_{7/2}$	43,831.08
$3s^2 6d$	$6d \ ^2\mathbf{D}_{3/2}$	44,166.48
	$6d \ ^2\mathbf{D}_{5/2}$	44,168.88
$3s^2 7s$	$7s \ ^2\mathbf{S}_{1/2}$	44,273.16
$3s^2 7p$	$7p \ ^2\mathbf{P}_{1/2}$	44,928.4
	$7p \ ^2\mathbf{P}_{3/2}$	44,930.4
$3s^2 6f$	$6f \ ^2\mathbf{F}_{5/2}$	45,194.65
	$6f \ ^2\mathbf{F}_{7/2}$	45,194.65
$3s^2 7d$	$7d \ ^2\mathbf{D}_{3/2}$	45,344.16
	$7d \ ^2\mathbf{D}_{5/2}$	45,345.60
$3s^2 8s$	$8s \ ^2\mathbf{S}_{1/2}$	45,457.27
$3s^2 7f$	$7f \ ^2\mathbf{F}_{5/2}$	46,016.73
	$7f \ ^2\mathbf{F}_{7/2}$	46,015.73
$3s^2 8d$	$8d \ ^2\mathbf{D}_{3/2}$	46,093.9

(continued)

Table 13.1 (continued)

Config	Term	Level (cm^{-1})
	8d ^{2}**D**$_{5/2}$	46,094.27
3s^{2}9s	9s ^{2}**S**$_{1/2}$	46,184.5
3s^{2}9d	9d ^{2}**D**$_{3/2}$	46,593.28
	9d ^{2}**D**$_{5/2}$	46,593.83
3s^{2}10s	10s ^{2}**S**$_{1/2}$	46,665.7
3s^{2}10d	10d ^{2}**D**$_{3/2}$	46,942.3
	10d ^{2}**D**$_{5/2}$	46,942.3

The ground state: 1s^{2}2s^{2}2p^{6}3s^2. The numerical values are extracted from a data book [1]

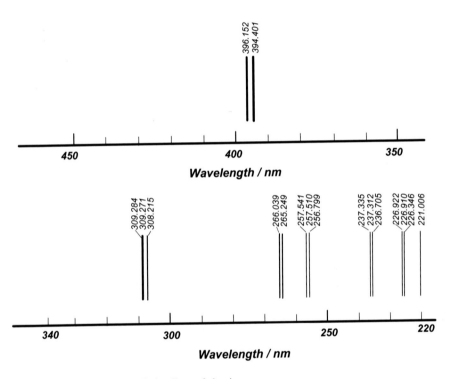

Fig. 13.2 Spectrograph of emission lines of aluminum

(4s) ^{2}**S**$_{1/2}$ → (3p) ^{2}**P**$_{1/2}$: $1/(25{,}347.69 - 0) = 394.51 \times 10^{-7}$ cm $= 394.51$ nm,
(4s) ^{2}**S**$_{1/2}$ → (3p) ^{2}**P**$_{3/2}$: $1/(25{,}347.69 - 112.04) = 396.26 \times 10^{-7}$ cm $= 396.26$ nm.
These spectral lines are observed at wavelength positions of 394.401 and 396.152 nm in the Al spectrum of Fig. 13.2. There is a slight difference between the calculated and the measured values. This difference is not because it is derived from a measurement error, but because the calculated values are in vacuum and the measured values are in the air medium.

Is it possible to cause electronic transitions from the second excited electron configuration, $(3s)(3p)^2\,{}^4\mathbf{P}_{1/2,3/2,5/2}$, to the ground-state electron configuration, $(3s)^2(3p)\,{}^2\mathbf{P}_{1/2,3/2}$? Such transitions are forbidden because of the selection rule for the spin quantum number, $\Delta S = -1$. In fact, spectral lines assigned to these transitions are not observed in the spectrum of the aluminum atom.

In the aluminum spectrum in Fig. 13.2, there are three spectral lines around 309 nm, at shorter wavelengths than 394 and 396 nm. These lines are attributed to electronic transitions from the excited electron configuration, $(3s)^2(3d)$, to the ground-state configuration, $(3s)^2(3p)\,{}^2\mathbf{P}_{1/2,3/2}$, due to the parity allowed. Four possible transition paths from this electron configuration are considered as follows:

$(3d)\,{}^2\mathbf{D}_{3/2} \to (3p)\,{}^2\mathbf{P}_{1/2}$: $1/(32{,}435.45 - 0) = 308.31 \times 10^{-7}$ cm $= 308.31$ nm,

$(3d)\,{}^2\mathbf{D}_{3/2} \to (3p)\,{}^2\mathbf{P}_{3/2}$: $1/(32{,}435.45 - 112.04) = 309.37 \times 10^{-7}$ cm $= 309.37$ nm,

$(3d)\,{}^2\mathbf{D}_{5/2} \to (3p)\,{}^2\mathbf{P}_{1/2}$ is forbidden and not observed because of $\Delta J = -2$,

$(4d)\,{}^2\mathbf{D}_{5/2} \to (3p)\,{}^2\mathbf{P}_{3/2}$: $1/(32{,}436.79 - 112.04) = 309.36 \times 10^{-7}$ cm $= 309.36$ nm.

Therefore, three spectral lines around 309 nm are assigned to the transitions between the $(3s)^2(3d)$ and the $(3s)^2(3p)$ energy levels. As found in Fig. 13.2, the aluminum spectrum includes other spectral lines in a wavelength range of 200–300 nm, which are assigned to electronic transitions from higher excited energy levels to the ground state energy level. Consider their transition paths between the energy levels in Table 13.1.

13.2 Atomic Spectrum of Calcium

Table 13.2 shows energy level tables of the neutral atom and the singly-ionized atom of calcium [1]. Figure 13.3 shows the atomic spectrum of calcium. What kind of electronic transitions cause the spectral lines in this spectrum? Here, in addition to the neutral atom, spectral lines of the singly-ionized atom of calcium are also observed.

Table 13.2 Energy levels for (a) neutral and (b) singly-ionized calcium atom

Config	Term	Level (cm^{-1})
(a) Ca I, the ground state: $1s^2 2s^2 2p^6 3s^2 3p^6 4s^2$		
$3p^6 4s^2$	$4s^2\,{}^1\mathbf{S}_0$	0.000
$3p^6 4s(^2S)4p$	$4p\,{}^3\mathbf{P}_0$	15,157.910
	$4p\,{}^3\mathbf{P}_1$	15,210.067
	$4p\,{}^3\mathbf{P}_2$	15,315.948
$3p^6 4s(^2S)3d$	$3d\,{}^3\mathbf{D}_1$	20,335.344
	$3d\,{}^3\mathbf{D}_2$	20,349.247

(continued)

Table 13.2　(continued)

Config	Term	Level (cm^{-1})
	3d $^3\mathbf{D}_3$	20,370.987
$3p^6 4s(^2S)3d$	3d $^1\mathbf{D}_2$	21,849.610
$3p^6 4s(^2S)4p$	4p $^1\mathbf{P}_1$	23,652.324
$3p^6 4s(^2S)5s$	5s $^3\mathbf{S}_1$	31,539.510
$3p^6 4s(^2S)5s$	5s $^1\mathbf{S}_0$	33,317.250
$3p^6 3d(^2D)4p$	4p $^3\mathbf{F}_2$	35,730.450
	4p $^3\mathbf{F}_3$	35,818.712
	4p $^3\mathbf{F}_4$	35,896.890
$3p^6 3d(^2D)4p$	4p $^1\mathbf{D}_2$	35,835.400
$3p^6 4s(^2S)5p$	5p $^3\mathbf{P}_0$	36,547.671
	5p $^3\mathbf{P}_1$	36,554.722
	5p $^3\mathbf{P}_2$	36,575.132
$3p^6 3d(^2D)4p$	4p $^1\mathbf{P}_1$	36,731.622
$3p^6 4s(^2S)4d$	4d $^1\mathbf{D}_2$	37,298.312
$3p^6 4s(^2S)4d$	4d $^3\mathbf{D}_1$	37,748.192
	4d $^3\mathbf{D}_2$	37,751.884
	4d $^3\mathbf{D}_3$	37,757.462
(b) Ca II, the ground state: $1s^2 2s^2 2p^6 3s^2 3p^6 4s$		
$3p^6 4s$	4s $^2\mathbf{S}_{1/2}$	0.000
$3p^6 3d$	3d $^2\mathbf{D}_{3/2}$	13,650.212
	3d $^2\mathbf{D}_{5/2}$	13,740.901
$3p^6 4p$	4p $^2\mathbf{P}_{1/2}$	25,191.541
	4p $^2\mathbf{P}_{3/2}$	25,414.437
$3p^6 5s$	5s $^2\mathbf{S}_{1/2}$	52,166.982
$3p^6 4d$	4d $^2\mathbf{D}_{3/2}$	56,389.309
	4d $^2\mathbf{D}_{5/2}$	56,858.511
$3p^6 5p$	5p $^2\mathbf{P}_{1/2}$	60,535.0
	5p $^2\mathbf{P}_{3/2}$	60,613.2
$3p^6 4f$	4f $^2\mathbf{F}_{5/2}$	68,056.86
	4f $^2\mathbf{F}_{7/2}$	68,056.95
$3p^6 6s$	6s $^2\mathbf{S}_{1/2}$	70,677.61
$3p^6 5d$	5d $^2\mathbf{D}_{3/2}$	72,722.11
	5d $^2\mathbf{D}_{5/2}$	72,730.77
$3p^6 6p$	6p $^2\mathbf{P}_{1/2}$	74,485.8
	6p $^2\mathbf{P}_{3/2}$	74,521.7
$3p^6 5f$	5f $^2\mathbf{F}_{5/2}$	78,027.8

(continued)

Table 13.2 (continued)

Config	Term	Level (cm^{-1})
	5f $^2\mathbf{F}_{7/2}$	78,027.8
3p^{6}5g	5g $^2\mathbf{G}_{7/2}$	78,163
	5g $^2\mathbf{G}_{9/2}$	78,163
3p^{6}7s	7s $^2\mathbf{S}_{1/2}$	79,449.9

The numerical values are extracted from a data book [1]

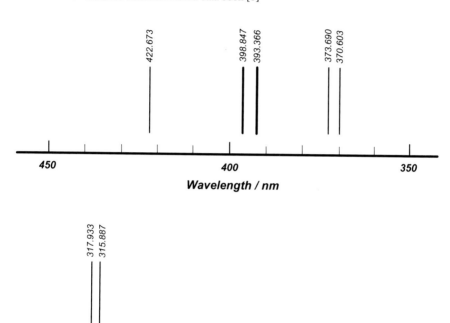

Fig. 13.3 Spectrograph of emission lines of calcium

Reference

1. Moore CE (1949) Atomic energy levels, vol 1. NBS Circular No. 467. U.S. Government Printing Office, Washington, DC

Chapter 14
Application of Atomic Emission Spectroscopy

14.1 Electric Discharge Plasma

How much energy is required to obtain the atomic spectrum? It is necessary to perform an electronic transition between energy levels of the outermost electron configuration. In an example shown in the previous chapter, an excited level of the aluminum atom is the (3d) $^2D_{3/2}$ level having an excitation energy of 32,435.45 cm^{-1}, which is converted into 32,435.45/8065 = 4.02 eV. Although there are some differences depending on the electron orbitals of analyte elements, their energies are within several eV, which are much smaller than the excitation for fluorescent X-rays described in Chap. 4. Therefore, it is relatively easy to obtain the atomic spectrum and to excite it by various methods.

In current analytical methods, electric discharge is employed as the excitation source for atomic emission spectrometry the most commonly. It is easily incorporated in an analytical apparatus because the electric discharge is a phenomenon that can be generated in our laboratory-scale conditions. For example, an electric spark in air atmosphere gives the atomic spectrum of atmospheric gases, such as nitrogen molecule and atom. When a high voltage is applied to a gaseous body, a part of it is ionized to cause the breakdown phenomenon in which the electrical conductivity is maintained. This is called *weakly-ionized gas* or "Discharge Plasma", which plays a role in exciting the atomic spectrum. What happens in the discharge plasma on an atomic scale? Electrons generated by the discharge are accelerated by the electric field to acquire kinetic energy. When the electrons collide with sample atoms, the kinetic energy is transferred so that they can be transited to certain excited states. The kinetic energy of electrons has a wide range from a few eV to several 100 eV, dependent on the type of the discharge. Since the number density of electrons is large, they have the ability to excite the sample atoms according to their energy distribution.

Radio-frequency inductively coupled plasma, which is abbreviated as ICP, is the most often used for atomic emission spectrometry [1]. Figure 14.1 is a schematic diagram of ICP. When a high frequency of 27.12 or 40.68 MHz is applied to the coil with a power of about 1 kW, a strong magnetic field is generated inside the coil due

© The Author(s), under exclusive license to Springer Nature Singapore Pte Ltd. 2021
K. Wagatsuma, *Spectroscopy for Materials Analysis*, SpringerBriefs in Materials,
https://doi.org/10.1007/978-981-16-5946-1_14

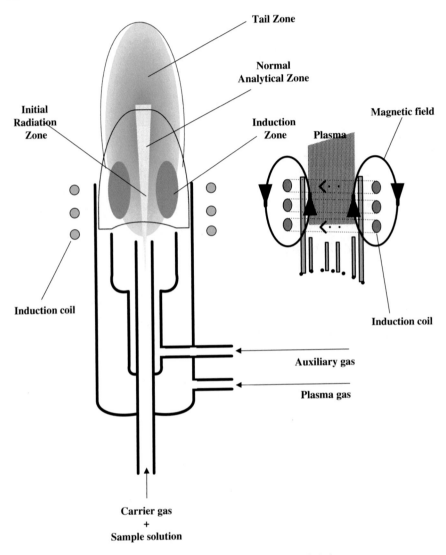

Fig. 14.1 Schematic structure of radio-frequency inductively-coupled plasma

to magnetic induction. At this time, if plasma gas (usually argon) is passed through a coaxially-arranged quartz tubes (plasma torch), the gas breaks down to form a plasma state. An illumining zone like a candle flame appears along the gas flow. When a sample aerosol is introduced into this plasma, the sample atoms are excited and emanate their characteristic emission lines which provide the qualitative/quantitative information. Since ICP can produce a plasma body with a large heat capacity, it is possible to directly introduce a sample solution, enabling the actual analysis to be conducted more easily. Furthermore, because the analytical zone of ICP is surrounded by a higher temperature zone (so-called inverse flame), ICP has a good analytical performance with the relatively low background continuum as well as the small self-absorption effect.

14.2 Spectrometer

The measuring apparatus for atomic emission spectrometry consists of discharge plasma as the excitation source, a spectrometer for dispersing the atomic spectrum by wavelength, and a detector that measures the intensity of each emission line. Spectrometers with several optical systems have been put into practical use, in which the most important one is a dispersing element (*monochromator*). The principle is exactly the same as the wavelength-dispersive measurement of X-rays using Bragg diffraction, as explained in Sect. 6.3. A *diffraction grating* is commonly used in atomic emission spectrometry because the wavelength of atomic spectra is several hundred nm, whereas analyzing crystals are used for X-ray diffraction because the wavelength of X-rays is several nm. The diffraction grating has several thousand grooves per mm on a metal substrate, which is suitable for the dispersion of electromagnetic wave in the ultraviolet/visible wavelength region.

Figure 14.2 shows a configuration of the measuring system in which an ICP is incorporated as the excitation source in atomic emission spectrometry. By aspirating a sample solution into the plasma, it is desolvated and then excited/ionized through the central channel of the plasma, and the spectrum of sample atoms is emitted. This system has a wavelength-scanning spectrometer to detect each of the emission lines by rotating the diffraction grating, which is employed the most commonly in commercial ICP apparatuses. A photomulitiplier tube (PMT) is generally used as the detector because it has a superior performance to other detectors with respect to the sensitivity as well as a linear response of the photo-current. As shown in Fig. 14.3, the structure and operation of PMT are analogous to a scintillation counter for X-ray detection (see Fig. 6.9), but it is not necessary to include a scintillation plate in PMT, since ultraviolet/visible radiation can be directly introduced into the photo cathode. Other than the scanning-type monochromator, polychromators such as Paschen-Runge and Echelle mountings [2], are also employed to realize multi-elemental and simultaneous determination in the ICP spectrometry. In such types of monochromator, an array of semiconductor detectors which are arranged lineally or two-dimensionally, such as a charge coupled device (CCD), is employed as the detector.

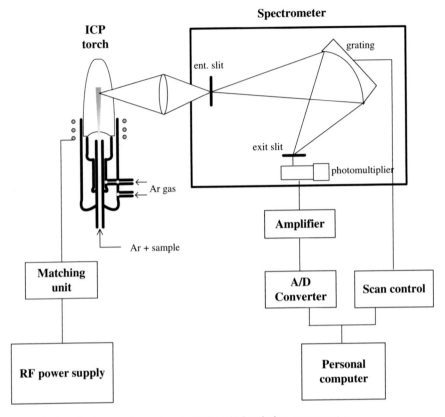

Fig. 14.2 A typical measuring system for ICP optical emission spectrometry

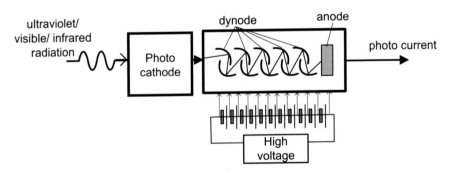

Fig. 14.3 Schematic diagram of photomultiplier tube

References

1. Boumans PWJM (1987) Inductively coupled plasma emission spectroscopy, part 1. Wily-Interscience, New York
2. Ingle JD Jr, Crouch SR (1988) Spectrochemical analysis. Prentice Hall, New Jersey

Chapter 15
Atomic Emission Spectrometry for On-site Analysis

15.1 What Kind of Spectrometric Method Is Suitable for the On-site Application?

In Chaps. 11–14, we see the principle of atomic spectroscopy based on an electron transition between outer electron energy levels and the excitation/de-excitation process to cause the transition. In fact, analytical methods using atomic emission spectroscopy are extensively adopted for obtaining elemental compositions in various samples and thus become a core technique in current quantitative analysis. Section 14.5 has explained radio-frequency inductively-coupled plasma (ICP) briefly, because its analytical performance can be suited to the precise and accurate determination of elemental compositions for a variety of analytes. Further, there are other types of the excitation sources generated by electric discharges, such as arc and spark discharges, and they are also employed as excitation/ionization sources in atomic emission spectrometry according to the characteristics of each source.

As already described in Sect. 1.5, on-site/on-line analysis occupies a particular position in the production process of various industrial materials. The most important issue for such analyses is a rapid response of the analytical result. For this purpose, several methods by which a solid sample is directly analyzed with little pretreatment are introduced at the production site. Here, we will focus on *glow discharge—optical emission spectrometry* (GD-OES) and *laser-induced plasma—optical emission spectrometry*. The latter is usually abbreviated as LIBS (originated from laser-induced breakdown phenomenon).

© The Author(s), under exclusive license to Springer Nature Singapore Pte Ltd. 2021
K. Wagatsuma, *Spectroscopy for Materials Analysis*, SpringerBriefs in Materials,
https://doi.org/10.1007/978-981-16-5946-1_15

15.2 Glow Discharge Optical Emission Spectrometry

15.2.1 Glow Discharge Plasma

Glow discharge is a steady-state discharge type that is generated in a reduced-pressure atmosphere of several hundred Pa. A self-stable steady discharge occurs according to Paschen's law, in which the type of discharge gas, the distance between electrodes, and the applied voltage are operated as the control parameters. In the voltage/current characteristics of direct-current (DC) discharge, the current for visible discharge is in a range of 1 mA–10 A [1]. The low current side of a few mA, where the load resistance becomes gradually reduced in an increase in the current, is called *normal glow discharge*, and further with increasing current, a bright discharge appears in a range of several tens of mA where the resistance is almost constant, which is called *abnormal glow discharge* [1]. The voltage/current region of the abnormal glow discharge is mainly used in glow discharge emission spectrometry. When the current further increases up to 1 A, the electric resistance of the discharge gas drops drastically, and then the discharge is changed to another form called *arc discharge*, which is also employed as an excitation source for atomic emission spectrometry.

Figure 15.1 schematically shows the structure of a type of *glow discharge plasma*. Unlike spark and arc discharges, the discharge voltage applied between the electrodes is localized and distributed in a dark portion near the surface of the cathode, and the portion is named as a *cathode dark space*. This type of the voltage distribution is known as *cathode drop*. In addition, there appears a bright emitting zone, called a *negative glow region*, outside the cathode dark space. In the glow discharge plasma, these two regions, the cathode dark space and the negative glow, mainly dominate the characteristics, which can explain most of the phenomena occurring in the plasma. Electrons are first supplied to the plasma through electron emission when plasma gas ions and/or the particles collide with the cathode surface (γ effect), while thermal electrons ejected by heating the cathode surface are much less. This effect is an essential difference from arc discharge and spark discharge plasmas, since the majority of electrons that maintain these discharges are supplied thermally. The glow discharge is sometimes called *cold cathode discharge* based on such characteristics.

The process for introducing sample atoms into the glow discharge plasma is generally called *cathode sputtering*. The process is that gas ions generated mainly in the negative glow are accelerated by the electric field in the cathode dark space and then collide with the sample surface, and that the kinetic energy released in the collision disturbs the atom arrangement near the sample surface and finally sample atoms of the outermost surface are ejected. The driving force is a large potential difference localized within the dark space of the cathode (cathode drop). A parameter that governs the amount of sample introduced is defined as the *sputtering yield*, which depends on the type of incident ions, the acceleration voltage, and so on, but does not directly correlate with the thermal properties of the sample. It is known that measured sputtering yields do not differ largely between the elements.

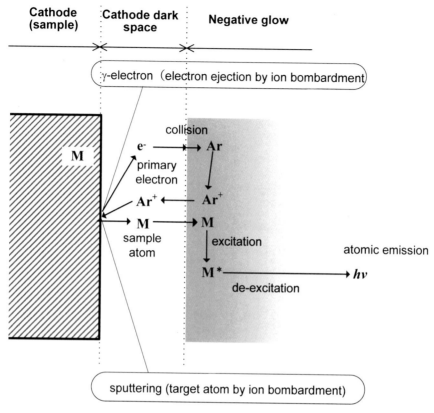

Fig. 15.1 Fundamental processes occurring in glow discharge plasma

Figure 15.2 schematically shows the behavior of electrons in glow discharge plasma [2]. The electrons (*primary electrons*) emitted from the surface of the cathode are accelerated toward the anode by the electric field in the dark portion of the cathode. Their kinetic energies become about the same as the discharge voltage (several hundreds of volts) through moving paths toward the anode, and the electrons repeatedly collide with gas atoms and then ionize them to generate *secondary electrons* in the plasma. Since it is a phenomenon under a reduced pressure, the collision frequency between gas particles is generally low, and further, since the high-speed primary electrons have a small collision cross section, most of them reach the anode without sufficient interactions with gas particles, which are called *run-away electrons*. In the negative glow region, the number density of positive ions becomes slightly higher than the density of electrons due to the difference in their mobility in the plasma. Therefore, the negative glow region exists between a sheath layer near the cathode with a large negative potential and another sheath layer near the anode because it has a small positive potential relative to the grounded anode. As shown in Fig. 15.2, this voltage difference is called a *plasma potential*. Because the

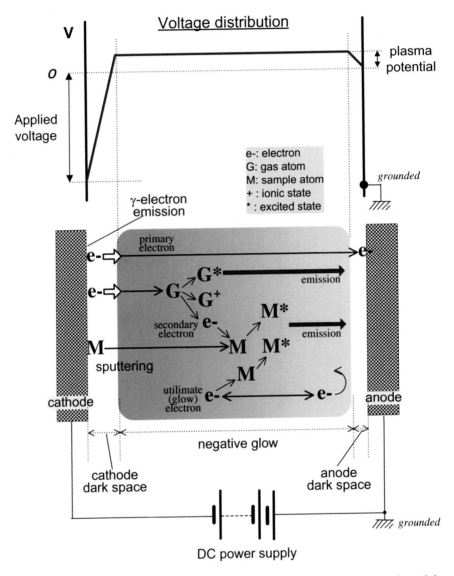

Fig. 15.2 Schematic voltage distribution between the electrodes of glow discharge tube and the behavior of electrons in the plasma

electric field gradient is small in the negative glow region, electrons that are generated in it are not greatly accelerated and their kinetic energy is small (about several eV). As a result, electrons in the glow discharge comprise three types of electrons having different energy distribution: the primary electrons that are greatly accelerated in the cathode dark space, the secondary electrons generated by collisions with gas particles in the plasma, and low-speed electrons that are held in the negative glow region by the plasma potential, called *ultimate or glow electrons* [3]. It is thought that the number density of the ultimate (glow) electron is much higher than the densities of the other electrons [3].

Since the glow discharge is maintained under a reduced pressure, the collision frequency between primary electrons and gas atoms is low, meaning that the energy distribution to neutral atoms and ions is very low. Namely, the gas temperature is significantly lower than the electron temperature. Here, the concept of temperature derives from the average kinetic energy of each particle. In electric discharges under an ambient pressure, *local thermodynamic equilibrium (LTE)* may be established between different kinds of particles due to a sufficient exchange in their kinetic energies [4]; that is, their average kinetic energies are almost the same in the plasma. However, it is difficult to fulfill the condition for the LTE relationship in the glow discharge plasma; in addition, the electron temperature is not uniform between the three kinds of electrons in the plasma. In this way, glow discharge generates a plasma state in which the LTE condition is not established and the constituent particles, such as electrons and gas atoms, have different characteristic temperatures. As a result, the observed spectral patterns and the intensities of emission lines are different from those in arc and spark discharge plasmas. Generally, the type of plasma gas in glow discharge plasma has a significant effect on its spectral characteristics, because various types of collisions with plasma gases as well as electrons contribute to the excitation of the emission spectrum while collisions with high-energy electrons principally determine the spectrum pattern in the other plasmas. The detailed mechanism for the plasma emission is described in Refs. [5–7].

15.2.2 Measurement Method

Several types of glow discharge tubes, having different shape and arrangement of electrodes, have been put into practical use. Among them, the excitation source that is widely used in atomic emission spectrometry is a hollow-anode and planner-cathode discharge tube developed by Grimm [8]. Figure 15.3 shows a schematic diagram of this discharge tube, indicating how the emitted radiation is observed from the plasma. When a sample plate (cathode) is set to a hollow anode having an inner diameter of 4–8 mm at an interval of 0.2–0.5 mm between the electrodes, a plasma gas of several hundred Pa is introduced inside the discharge tube and then a DC voltage of several hundred V is applied to produce glow discharge plasma. This plasma is categorized into an *obstructed glow discharge* such that the negative glow region can be localized in front of the sample electrode [1], and thus the sample atoms are introduced convergently into this region and emits the intense spectrum. The plasma is extremely stable, thus enabling highly accurate quantitative analysis. A flat-shaped solid sample can be analyzed with a simple operation, which is suitable for on-site analysis at the production site in materials industry.

Figure 15.4 shows a block diagram of the measuring device in GD-OES. It consists of a glow discharge tube as the excitation source, a power supply, a monochromator,

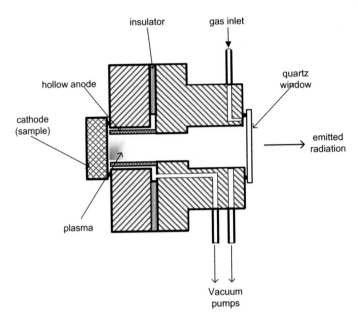

Fig. 15.3 Cross-sectional structure of a Grimm-type glow discharge lamp

and a detector and a data processing device. In commercial apparatuses, a Paschen–Runge type monochromator is widely used in the spectrometer system. This has an optical arrangement also called a *polychromator* which can focus images of the spectral lines on one circle (*Roland circle*) by a curved diffraction grating, in which exit slits combined with photomultiplier tubes are fixedly arranged at the wavelength positions to be measured. Unlike wavelength-scanning spectrometers, this measuring apparatus can simultaneously record a number of spectral lines; accordingly, it is suitable for multi-elemental quantification and tracking of a transient phenomenon at high speed. A DC power source is usually used for the glow discharge, but a method using a radio-frequency power source is also possible, and in this case, it can also be applied to the analysis of electrically insulating samples.

15.2.3 Analytical Application

GD-OES is employed for the accurate determination of constituent elements in metallic materials. Figure 15.5 shows a calibration relationship between the intensity of a phosphorous emission line and the content in steel materials [9]. Phosphorous remains as an impurity element in the materials and may exert a negative effect in the forming process of them; therefore, the content should be strictly controlled through on-site analysis in the steel-making process. Fluctuations in the emission intensity

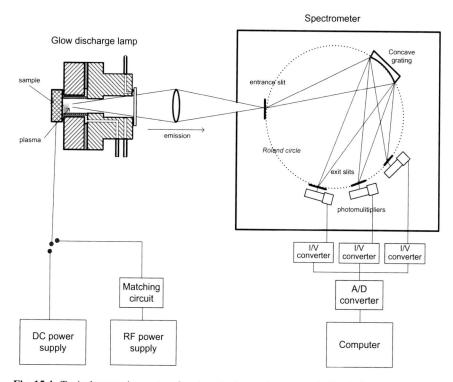

Fig. 15.4 Typical measuring system for glow discharge plasma—optical emission spectrometry

are estimated using the *relative standard deviation* (RSD), which is the standard deviation divided by the average value. The RSDs are less than 1.6% in the measurement of Fig. 15.5, which is generally superior to other spectrometric techniques.

GD-OES has the characteristics suitable for direct analysis of solid samples as an analytical method that enables rapid, simultaneous, and multi-elemental determination, and can detect most of elements except rare gases. Moreover, since the sample atoms are introduced into the plasma by cathode sputtering, there is a unique analytical feature that variations in the chemical composition can be measured in the depth direction, which is called a *depth profile*. Especially in the on-site analysis, various types of industrial products, such as an electroplated layer on steel materials, are depth-profiled using GD-OES for the quality control. Figure 15.6 presents a typical example of the depth profiling, where a Zn electroplated layer on a steel substrate, having a duplicate structure of different chemical composition, is measured from the surface into the substrate. In this respect, GD-OES can be classified into a surface analysis method. As described in Sect. 10.1, XPS and AES combined with ion etching may provide a similar depth profile; however, GD-OES has an advantage over these electron spectroscopies because it can be carried out rapidly with simpler operations.

Fig. 15.5 Calibration curve
of the emission intensities of
PI 177.49 nm and their
relative standard deviation
when phosphorous in steel
samples is quantified in
GD-OES. This article is
reprinted from [9] under the
permission of The Japan
Society for Analytical
Chemistry

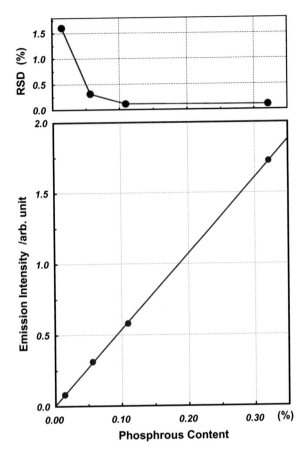

Fig. 15.6 Changes in the
emission intensity of Fe and
Zn emission lines as a
function of the sputtering
time (depth profile) for a
duplicate Zn/Fe coating on a
steel substrate. The nominal
thickness of the coated layer
is about 4 μm

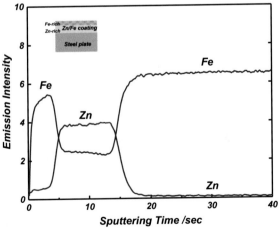

15.3 Laser-Induced Plasma—Optical Emission Spectrometry

15.3.1 Laser-Induced Breakdown Plasma

We will see the characteristics of the laser-induced breakdown phenomenon. When a solid surface is irradiated with a laser beam having a high energy density, a large number of electrons are supplied to the surrounding gas near the surface and electron collisions occur as a chain reaction, so that a part of the gas atoms/molecules can be ionized. It becomes *weakly-ionized gas*. This phenomenon is called *laser breakdown* and triggers the plasma generation. At the same time, the thermal evaporation of sample atoms occurs on the sample surface due to the incident energy of the laser beam, and eventually the sample atoms are ejected towards the plasma. This phenomenon is called *laser ablation* and is used for sampling of analyte atoms in the spectrometric method.

Figure 15.7 schematically shows a transient phenomenon in the plasma induced by a single laser shot. High-energy electrons, which are emanated from the sample surface immediately after the breakdown, repeatedly collide with gas particles and expands towards the surrounding gas. The behavior of the electrons is quite different from that in the glow discharge plasma, because they are not accelerated in the laser-induced plasma and thus are quenched rapidly to lead to extinction of the plasma. Since the sample atoms undergo various excitation processes in this region as well, the emission spectrum varies spatially and temporally due to the creation/extinction of the pulsed plasma. As shown in Fig. 15.7, the emission intensity of the sample atom changes with the progress of the plasma, and is generally expected to reach a maximum value immediately after the plasma generation. This is due to the fact that

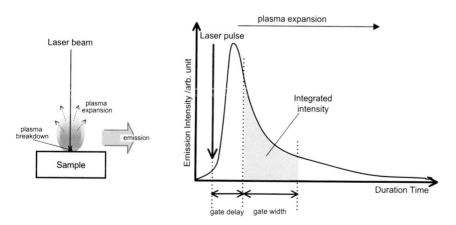

Fig. 15.7 Temporal change of laser-induced breakdown plasma and the transient emission signal from the plasma (schematic drawings)

a large amount of high-energy particles are generated in a high-density gaseous body in the early stage of the plasma and actively repeat excitation collisions with sample atoms. However, since there are gas particles that cause the excitation collision even during the expansion period of the plasma, the emission spectrum of the sample atom is still observed [10, 11]. In the LIBS method, it does not necessarily provide a good analytical result that the time-dependent emission intensity is integrated totally over the entire transient region. This is because, immediately after the breakdown, a large number of high-speed electrons cause bremsstrahlung when they collide with gas particles. This effect contributes to high background emission that interferes with the measurement of the emission intensity from sample atoms. Further, in such a high-density plasma body, it may be problems that the emission spectrum is degraded by the pressure broadening of the line profile and the self-absorption phenomenon.

While the time-integrated measurement is usually carried out in atomic emission spectrometry, a time-resolved measurement method is recommended in the LIBS method. In this case, the important experimental factors are a delay time for measurement and a gate width in the measurement, as indicated in the right figure of Fig. 15.7. The *signal-to-background ratio* (SBR), the *signal-to-noise ratio* (SNR), etc. are examined to optimize the measuring conditions for each sample and each emission line. Figure 15.8 shows a variation in the intensity of an Al emission line as a function of the gate delay time in LIBS [12]. The emission intensities generally decrease with an increase in the gate delay time. However, their fluctuations, which are presented with error bars in Fig. 15.8, take relatively large values when the delay time is shorter than 400 ns, but the values rapidly decrease at delay times of longer than 500 ns, improving the measurement precision. This is because the background emission due to recombination and Bremsstrahlung radiation of gas species, becomes intense just after the laser breakdown. Therefore, the gate delay time is optimized to be 500 ns in this measurement.

Fig. 15.8 Variation of the emission intensity of Al I 309.271 nm as a function of the gate delay time with a fixed gate width of 500 ns. This article is reprinted from [12] under the permission of The Japan Society for Analytical Chemistry

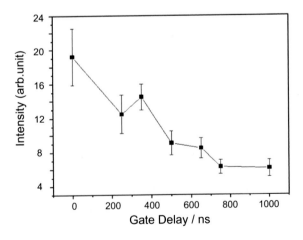

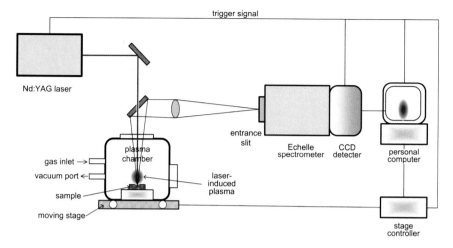

Fig. 15.9 Setup of the measurement apparatuses for laser-induced plasma—optical emission spectrometry

15.3.2 Measurement Method

Figure 15.9 illustrates a block diagram of the LIBS measurement system employed in commercial apparatuses. It consists of a high-power laser, a sample chamber, a monochromator, and a detector and the data processing device. The most popular laser source is a *Q-switched Nd:YAG laser* in a pulse operation. Since the repetition frequency of the laser is generally several ten Hz, the laser irradiation on the sample surface is performed in the form of discontinuous pulses, and the resultant plasma is also generated discontinuously. It is generally known that, when a Nd:YAG laser is operated with a pulse width of about 10 ns, the emission signal can be obtained with good SBR and SNR when the delay time is set to several hundred ns and then the measurement is started. In this selection, the type of the surrounding gas and the gas pressure are also important experimental parameters. These results indicate that an analytical emission line should be measured with a low background intensity during the expansion period of the plasma when the high background emission almost disappears. For this purpose, we need a spectrometer system such that a time-resolved measurement can be conducted and that a spectrum can be recorded over a wide wavelength range at a fast response. A combination of an *Echelle monochromator* and an *ICCD detector* is suitable for these measuring conditions.

15.3.3 Analytical Application

LIBS has several features suitable for the direct analysis of solid samples, such as on-site/in-line analysis for industrial use, in which an as-received specimen can be

Fig. 15.10 Calibration curve for the emission intensity of Mg I 285.211 nm in aluminum alloy samples in the LIBS measurement. This article is reprinted from [12] under the permission of The Japan Society for Analytical Chemistry

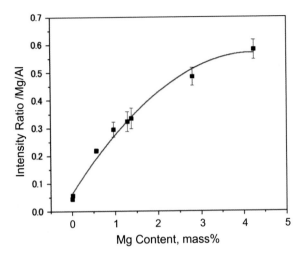

analyzed with little pre-treatment without any direct contact with it. Accordingly, the analytical procedure can be conducted easily and rapidly.

The qualification in the LIBS measurement is commonly carried out through a method in which calibration curves are estimated using several standard samples. A typical example for the calibration method is shown in Fig. 15.10 [12], where the intensity of an Mg emission line is plotted for different contents of magnesium in several aluminum alloy samples. The curve deviates from a linear relationship at higher contents of magnesium due to the self-absorption effect, and their variations are relatively large compared to GD-OES (see Fig. 15.5). This is a limitation of LIBS because of the pulse-like plasma.

Quantitative and morphological evaluation of non-metallic inclusion particles, which are embedded in the iron matrix, is an important issue in the analysis of steel products, because they may degrade the quality of the steel product. Conventional analytical methods for estimating such a distribution are based on a direct observation by using an optical microscope; however, this method requires time-consuming and complicated procedures for the measurement as well as the data handling. LIBS can provide the spatial distribution of such inclusions so rapidly. Figure 15.11 indicates an application of the elemental mapping by the LIBS measurement [13]. A two-dimensional distribution of alumina inclusion particles in ferritic stainless steels is determined from variations in the emission intensity of an Al emission line, when a 1-kHz Q-switched Nd:YAG laser, having a beam diameter of about 50 μm, is scanned on the sample surface and the emission intensity is simultaneously recorded. A histogram for detected alumina particles can be estimated as a function of the averaged diameter or the emission intensity of the particles. This technique has an advantage over conventional detection methods, because the scanning LIBS provides the analytical result instantly whereas the conventional ones require prolonged analysis time.

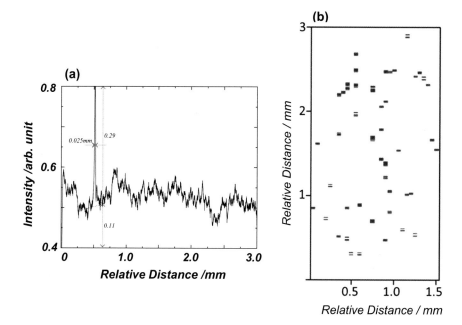

Fig. 15.11 Linear-directional (**a**) and two-dimensional (**b**) distribution of alumina inclusion particles precipitated in a stainless steel sample which are estimated from changes of the emission intensity of Al I 396.142 nm in the LIBS measurement. A 1-kHz Q-switched Nd:YAG laser is irradiated on the sample surface with an irradiation power of 1 mJ/pulse and a moving rate of the laser of 0.5 mm/s. This article is reprinted from [13] under the permission of Elsevier

References

1. von Engel A (1965) Ionized gases. Clarendon Press, Oxford
2. Wagatsuma K (1998) High Temp Mater Process 17:97–116
3. Chapman A (1980) Glow discharge processes. Wiley, New York
4. Boumans PWJM (1966) Theory of spectrochemical excitation. Plenum Press, New York
5. Steers BM, Fielding RJ (1987) J Anal At Spectrom 2:239–244
6. Wagatsuma K, Hirokawa K (1991) Spectrochim Acta Part B 46:269–281
7. Wagatsuma K, Honda H (2005) Spectrochim Acta Part B 60:1538–1544
8. Grimm W (1968) Spectrochim Acta 23B:443–454
9. Wagatsuma K (2000) ISIJ Int 40:783–788
10. Kitaoka C, Wagatsuma K (2007) Anal Sci 23:1261–1265
11. Weng J, Kashiwakura S, Wagatsuma K (2020) ChemistrySelect 5:12558–12563
12. Fugane Y, Kashiwakura S, Wagatsuma K (2020) Anal Sci 36:1415–1421
13. Matsuda T, Kashiwakura S, Wagatsuma K (2020) Microchem J 153:104400

Printed in the United States
by Baker & Taylor Publisher Services